Y0-BYK-731

Sustainable Production Consumption Systems

Louis Lebel • Sylvia Lorek • Rajesh Daniel
Editors

Sustainable Production Consumption Systems

Knowledge, Engagement and Practice

Editors
Dr. Louis Lebel
Faculty of Social Sciences
Unit for Social and Environmental Research
Chiang Mai University
Thailand

Dr. Sylvia Lorek
Sustainable Europe Research Institute
Overath
Germany

Rajesh Daniel
Faculty of Social Sciences
Unit for Social and Environmental Research
Chiang Mai University
Thailand

ISBN 978-90-481-3089-4 e-ISBN 978-90-481-3090-0
DOI 10.1007/978-90-481-3090-0
Springer Dordrecht Heidelberg London New York

Library of Congress Control Number: 2009930634

Printed on acid-free paper

Springer is part of Springer Science+Business Media (www.springer.com)

Contents

Contributors

Rajesh Daniel
Unit for Social and Environmental Research, Chiang Mai University,
Chiang Mai, Thailand
rajesh@sea-user.org

Doris Fuchs
Professor of International Relations and Development, University of Münster,
Münster, Germany
doris.fuchs@uni-muenster.de

Po Garden
Internews' Earth Journalism Network, Chiang Mai, Thailand
Po@internews.org

Shabbir H Gheewala
Joint Graduate School of Energy and Environment (JGSEE), King Mongkut's
University of Technology, Bangkok, Thailand
shabbir_g@jgsee.kmutt.ac.th

Dao Huy Giap
Department of Agriculture, Hanoi University of Agriculture, Hanoi, Vietnam
awuaasia@gmail.com

Agni Kalfagianni
Department of International Relations and European Integration,
University of Stuttgart, Stuttgart, Germany
agni.kalfagianni@sowi.uni-stuttgart.de

Louis Lebel
Unit for Social and Environmental Research, Chiang Mai University,
Chiang Mai, Thailand
louis@sea-user.org

Giok Ling Ooi
National Institute of Education, Nanyang Technological University, Singapore
giokling.ooi@nie.edu.sg

Sylvia Lorek
Sustainable Europe Research Institute (SERI), Head of Sustainable Consumption Research, Overath, Germany
sylvia.lorek@seri.de

Michael Maniates
Professor of Political Science and Environmental Science, Allegheny College, Meadville, PA, USA
michael.maniates@allegheny.edu

Ligia Noronha
Director of the Resources and Global Security Division, The Energy and Resources Institute (TERI), New Delhi, India
ligian@teri.res.in

Benjamin Nölting
Center for Technology and Society, Berlin Institute of Technology, Berlin, Germany
noelting@ztg.tu-berlin.de

Fritz Reusswig
Potsdam Institute for Climate Impact Research, 14412 Potsdam, Germany
fritz@pik-potsdam.de

Hannarong Shamsub
RMIT International University, Ho Chi Minh City, Vietnam
hannarong.shamsub@rmit.edu.vn

Arnold Tukker
Sustainable Innovation, TNO, Delft, Netherlands
arnold.tukker@tno.nl

Uchita de Zoysa
Executive Director, Centre for Environment & Development (CED)
uchita@sltnet.lk

Chapter 1
Production–Consumption Systems and the Pursuit of Sustainability

Louis Lebel and Sylvia Lorek

Introduction

The pursuit of sustainability has many facets. Sectoral approaches attempt to improve the productivity of agriculture or energy efficiency while reducing negative impacts of air and water pollutants on the environment. In place-based approaches a suite of environmental challenges posed by development are tackled together seeking to reduce underlying drivers, complementarities among inputs and inputs, and negotiating trade-offs when win-win situations are hard to find. In product-oriented approaches the focus is on reducing material and energy or need for hazardous or environmentally threatening compounds used in the manufacture of a particular product. Consumer-oriented approaches use information campaigns to attempt to change what people buy or how they use particular services or goods to lessen impacts on the environment.

Each of these approaches to sustainability has its limitations. Too narrow a focus on energy security and sustainability may mean for instance, ignoring the impacts of the expansion of agrofuels on other sectors like agriculture and food. Efforts to tackle environmental problems in one city, while successful, may ultimately just end up shifting problems of polluting industries or waste disposal to another place. Gains in fuel efficiency of cars may reduce pollution loads per kilometre travelled but be offset by households travelling more and farther or buying a second car and making separate trips. Too much competing information on how to buy to "save the planet" creates confusion, may neglect how people behave towards products, and ignores options that don't involve purchasing, like sharing or *not* buying.

The pursuit of sustainability cuts across wealth classes. Over-consumption, while not providing significant benefits to improved well-being, is environmentally damaging (Hamilton and Denniss 2005). Under-consumption, while making individual

L. Lebel
Unit for Social and Environmental Research, Chiang Mai University, Chiang Mai, Thailand
e-mail: louis@sea-user.org

S. Lorek
Sustainable European Research Institute, Overath, Germany
e-mail: sylvia.lorek@seri.de

L. Lebel et al. (eds.), *Sustainable Production Consumption Systems: Knowledge, Engagement and Practice*, DOI 10.1007/978-90-481-3090-0_1,

lives miserable, hinders asset accumulation, market development and innovation in societies. Developing country concerns with maintaining soils and growing enough food, and controlling pollution from nascent but rapidly growing industries, are not separable or disconnected from the actions in industrialized or service economies. Investment, trade, production and consumption activities increasingly link across national boundaries. Sustainable consumption movements framed from a western wealthy perspective may undermine livelihoods and firms in developing countries. Neat separations into developing and developed, or south and north, ignore huge differences in access to and power to control critical resources within those categories and cross-linkages. Consumption with environmentally significant impacts (Stern et al. 1997) takes place through the activities of a global consumer class in the capitals of the developing world (Myers and Kent 2003) while the poor may be homeless in the cities of the developed. Gross over-consumption and acute under-consumption co-exist in the real world (Kates 2000) and both must be dealt with in the pursuit of sustainability (Narayan et al. 2000; Hamilton 2003).

The pursuit of sustainability is never-ending. Environmental conditions and patterns of resource use are dynamic and interact in complex ways. As this book was being completed many national economies were entering a recession as a result of the collapse of various inter-linked global financing schemes. Many developing countries were facing prospects of sharp contractions in exports as local manufacturing and agro-industries slowed production in response to drops in global demand. Unemployment was rising and consumers were cutting back on non-essential expenditures. During the same period public concerns and thus policy attention about climate change was reaching new heights. For the first time in many countries policy-makers were starting to think through the difficulties of financing and carrying out adaptation recognizing that in rapidly developing countries vulnerabilities and capabilities were shifting rapidly. Sustainability is forever being pursued in uncertain, complex and dynamic settings.

The pursuit of sustainability is knowledge intensive. Science could play a larger role than it has so far (Kates and Clark 1999; WSSD 2002). But to contribute effectively better ways are needed to bring together research-based understanding with the experience-based knowledge embedded in diverse local practices (Cash et al. 2003). Practice and knowledge often need to be brought closer together, but doing so requires going beyond a "pipeline model" of science to technology development to practice (van Kerkhoff and Lebel 2006). The engagement between science and practice or policy can vary from participation as consultation through to negotiation and mutual learning relationships.

Actions and analyses are needed which consider production and consumption together. We need to examine both supply and demand sides of relationships in systems of exchange that are dynamic and include both over- and under-consumption. We need to look, not only at the two ends of the chain, but also what goes on in-between as commodities stretch across the globe. We need to understand how knowledge and information influence the behaviour of governments, firms, households and individuals. In short, we should focus on Production–Consumption Systems (PCS) which we define in this book as systems in which environmental goods and services, individuals, households, organizations and states are linked by flows of energy and materials and relationships in which transactions of money and information or negotiation of power and influence take place (Lebel and Lorek 2008). Such approaches would not replace the vast set of experiences and efforts with sector-, place-, product- and consumer-oriented approaches to sustainability, but they would provide an important complement offering a more integrative and systemic view (Princen et al. 2002; Lebel 2004).

Pursuits

There are several reasons why a PCS approach to sustainability challenges is desirable.

Firstly, a PCS perspective brings attention to the processes closer to the decisions and actions of final consumers that help drive demand, and hence, are an underlying reason for environmental impacts at remote "production" parts of commodity chains (Hertwich 2005). Our understanding of consumers is increasingly being linked to environmental impacts and questions of sustainability (Stern et al. 1997; Hertwich 2002; Hobson 2003a). This body of research underlines the importance of housing, food and mobility as priority fields of action at the household level (Lorek and Spangenberg 2001; Spangenberg and Lorek 2002; Tukker and Jansen 2006).

Secondly, the commodity chain itself can be thought of as a series or network of many production and consumption relationships (Princen et al. 2002). For each linkage we can ask questions from both a production perspective (how could this industrial process be made more resource efficient?), and, in addition, a consumption perspective (what are the underlying drivers of demand?).

Thirdly, leverage to improve performance in the direction of sustainability can exist at diverse locations along chains sometimes remote from where pollution is emitted or a resource over-harvested. Many problems related to consumption do not result directly from dangerous and inefficient production processes (Murphy 2001). Likewise a narrow focus on behaviour of the household consumer may fail to identify much more powerful leverage points to reduce environmental impacts by meeting a need or aspiration in a different way (Meadows 1999; Clay et al. 2005). A PCS perspective allows consideration of alternative leverage points and the possibility of targeting specific demand or supply processes where necessary (Lebel 2004).

Finally, a PCS perspective is inherently interdisciplinary, taking into account material and energy flows common to life cycle and related methods but also embracing sociological insights from exploring behaviour and relationships among actors.

In short a PCS perspective could lead to strategic waypoints towards achieving sustainable development. A focus on sustainable production and consumption systems reduces the problem of pseudo-sustainability that arises from the shifting of environmental burdens onto others in the interest of a clean backyard at home. It also draws attention to the opportunity of negotiating shared responsibilities for action across disparate parts of linked value chains among firms and consumers.

The sustainability of production and consumption is already well recognized as central to achieving sustainable development by world leaders (Manoochehri 2002; Hertwich 2005). In the decade since the formulation of Agenda 21 at the Earth Summit in Rio de Janeiro (United Nations 1992) their importance has been reiterated at international meetings: at the World Summit on Sustainable Development in 2002 (WSSD 2002; Barber 2003) and more recently as part of the "Marrakesh Process" led by United Nations Environment Program (UNEP) and UN Department of Economic and Social Affairs (Clark 2007).

There is now a growing body of scholarship exploring a diverse range of initiatives and experiments aimed at enabling sustainable PCS (Lebel and Lorek 2008). Although the diversity of enabling mechanisms is large, it can be organized into discrete classes as shown in Table 1.1.

The 11 enabling mechanisms identified in Table 1.1 differ in which activities they target for intervention and who is expected to act or change practices. They also differ in the mixture of assumptions they make about conditions or instruments with respect to market institutions, government regulation, socio-technical innovation, and actor partnerships. Other classes of enabling mechanisms undoubtedly also exist.

A PCS perspective offers possibilities for measurement and assessment based on flow of materials and energy as well as relationships of actors and the distribution of values. Conventional product-oriented life-cycle analyses techniques, for instance, are suitable for describing resource inputs, waste outputs and energy flows of specific production processes and goods. In combination with input-output analyses and information about expenditure these tools are also being adapted for looking at baskets of goods or services essential for strategic analysis of sustainable consumption (Hertwich 2005). Commodity chain analyses suggest possibilities for developing metrics and indicators of how both benefits and involuntary risks are distributed among different actors which, interpreted through political economy lens, provides a means for exploring systematically issues of power. Social accounting matrices may help with deriving and exploring distributional issues, for example, across household types (Duchin and Hubacek 2003). For PCS to be a useful unit of analysis a more explicit conceptualization of the "system of interest" will usually be needed, and if comparisons are to be made, these will need to be done in reasonably consistent ways. In Box 1.1 we offer some suggestions.

Gaps

A recurrent challenge in pursuits of sustainable PCS (Box 1.1) is gaps between knowledge and practice. In some cases knowledge appears to be insufficient or lack saliency, whereas in others knowledge appears adequate but other factors shape

Table 1.1 Examples of enabling mechanisms for sustainable production–consumption systems (after Lebel and Lorek 2008)

Enabling mechanism	Short description	Concerns, constraints, or challenges
Produce with less	Innovations in production process reduce the environmental impact per unit made.	Rebound effects occur that wipe out gains through increases in the number of units or how they are used.
Green supply chains	Firms with leverage in a chain impose standards on their suppliers to improve environmental performance.	There may be unfair control of small producers.
Co-design	Consumers are involved in design of products and services to fulfil needs with less environmental impact.	Incentives are not adequate to involve consumers.
Produce responsibly	Producers are made responsible for waste from product disposal at end of life.	Incentives for compliance without regulation may be too low for many types of products.
Service rather than sell	Producers provide service rather than sell or transfer ownership of assets, which reduces number of units made while still providing functions needed.	This is a difficult transition for firms and consumers to make as it requires new behaviors and values.
Certify and label	Consumers preferentially buy labeled products. Labels are based on independent certification, and producers with good practices increase their market share.	Consumers are easily confused with too much information or with a lack of transparency and credibility of competing schemes.
Trade fairly	Agreements that may include a minimum price and other investments or benefits are made with producers. Consumers preferentially buy products labeled as or sold through fair trade channels, and producers get a better deal.	Mainstream trade still dominates. It is hard to maintain fair trade benefits to producers when a product becomes mainstream.
Market ethically	Reducing unethical practices in marketing and advertising would reduce wasteful and over-consumption practices.	Policy makers are reluctant to make or enforce regulation to tackle very powerful private sector interests.
Buy responsibly	Campaigns educate consumers about impacts of individual products, classes of products, and consumption patterns, resulting in overall behavior changes.	Converting intentions and values into actions in everyday life is often difficult for consumers. Issues of price, convenience, flexibility, and function still matter a lot.

(continued)

Table 1.1 (continued)

Enabling mechanism	Short description	Concerns, constraints, or challenges
Use less	Consumption may be reduced for a variety of reasons, for example, as a consequence of working less. There are many potential environmental gains from less overall consumption.	Difficult to get past the dominant perception that using or buying less means a sacrifice. Also, less income and consumption may not automatically translate into "better" consumption.
Increase wisely	Increasing the consumption of those who under-consume can be effected in ways that minimize environmental impacts as economic activity expands.	Wealthy developed countries need incentives and goodwill to assist the poor in developing countries, for example, by leaving adequate natural resources and ecosystem services for them to develop.

Box 1.1 PCS

In production–consumption systems (PCS) environmental goods and services, individuals, households, organizations and states are linked by flows of energy and materials and relationships in which transactions of money, information and influence take place (Lebel and Lorek 2008). A generalized view of material flows and derived utilities is given in Fig. 1.1. Production (P) transforms inputs from the harvest of environmental resource (E) into a good or service, which, in turn, is provided to consumers (C). Post-consumption materials and energy may be reused, recycled or returned to renew the environment. Otherwise it becomes waste. Postproduction materials may be

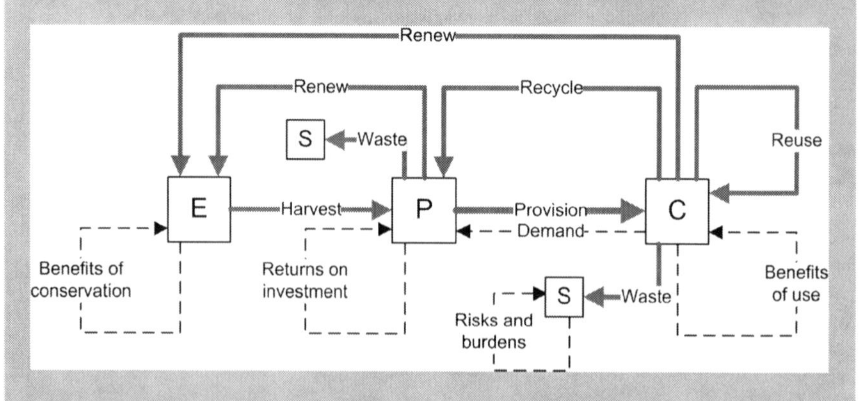

Fig. 1.1 A generalized production-consumption system (after Lebel and Lorek 2008)

Box 1.1 (continued)

returned to ecosystems for renewal or become waste. Waste accumulates in sinks (S), which can degrade the environment and affect people, including consumers and producers (e.g., employees). Each closed link in a commodity network or value chain is both a production and a consumption relation.

Each action, including producing and consuming, brings both utility and risk, taken and involuntary, for the actor themselves and for other actors as well. For example retailing brings receipts to the store that sells a good to a consumer, and the consumer benefits from possessing the good including an increased social status. Utility is derived from a PCS in several distinct ways: benefits of use, returns on investment, and benefits of conservation (Fig. 1.1). A fourth utility is negative: the risks and burdens associated with sinks.

Different flows and links are associated with different actors or roles. An actor may play more than one role. For example a large firm may have its own logistics capacity to distribute products and create demand through marketing. A subsistence farming household may be both producer and consumer and take on many of the other roles shown here. The concentration of multiple roles in a single actor may make coordinated change, and resistance to change, easier. Some roles more or less match material and energy flows like *recycle*, whereas others may operate relatively independently, for example, to *regulate* or *innovate*.

Individual actors in each role may vary substantially in characteristics. Households may be wealthy or poor, firms big or small, government departments powerful or weak. Heterogeneity of actors could help the stability of PCS if uncertainty and variability are common features of the resource base. Of course actor heterogeneity is also a likely outcome of differential success in those roles for both inherent reasons (effort, skill) and the magnification of discrimination and other forms of injustice.

Actors, including producers and consumers, are bound by institutional relationships that shape their practices and which their past behavior has itself helped construct (Hobson 2003b; Spaargaren 2003). The dynamics of PCS are likely to be affected by the social structures in which actors do their things. Individual practices, for example, may also be linked through non-consumption actions like boycotting the products of a company with a poor environmental performance.

Values and beliefs play an important role in enabling such social movements. Firms on the other hand may use appeals to sustainable practices in their labeling and marketing and gain a business advantage. These practices can become part of corporate culture. Shared norms and principles may govern both supply and demand (Lorek and Lucas 2003). Not much is known about value changes and transfers among societies interacting through production and consumption linkages, but it is clear there are some major differences as well as similarities across nations and over time (Leiserowitz et al. 2006).

Box 1.1 (continued)

Finally, the attributes of ecosystems and environments (E, Fig. 1.1) are important for the sustainability of production and consumption systems. Some resources are amenable to exclusive use whereas others are not (Ostrom 2003). Some natural resources are substitutable now or potentially substitutable in the future, whereas others are unlikely to be. Resilience, a key property of ecosystems, enhances the capacity to regenerate goods and services even as ecosystems are disturbed or altered (Gunderson 2000). Ecosystems are not static. They frame longer term changes in production and consumption systems. Consider for example the evolution of rice cultivation landscapes in the major deltas of Southeast Asia. These regimes have emerged over centuries involving mechanisms both to maintain social order as well as transform the landscape and ecosystems upon which, eventually, dense urbanized civilizations could eventually be built.

A PCS is considered sustainable if the transformation of energy, materials, knowledge and relationships maintains or improves human well-being and capacity to adapt without (a) reducing the long-term ability of ecosystems to provide a diverse set of goods and services; and (b) unfairly placing some people at disadvantage or exposing them to greater involuntary risk.

In practical terms comparing and assessing changes over-time in the performance of PCS will benefit from having metrics at least to cover those aspects of relationships amenable to partial quantification. Other aspects may benefit from checklists and more in-depth description of social processes, linkages and their changes.

decisions or practices. Gaps between knowledge and action arise and are maintained for three reasons. First, the specific knowledge needed may not be available to the relevant actors either because it doesn't yet exist or it is not being shared. Second, relevant actors may know what to do but do not have the capacity or power to act. And, third, decision-making may be dominated by criteria other than those of knowledge about sustainability.

A conventional and widely held view is that gaps between knowledge and practice can be closed with better communication, so that scientists and experts can tell practitioners and decision-makers what they need to know. Research on knowledge systems for sustainable development frequently suggests otherwise (Cash et al. 2003, 2006; van Kerkhoff and Lebel 2006). Gaps may be caused by problems aside from communication, including distinct problem framings, interest-based differences in priorities and political agendas, modes of engagement and other reasons. The important gaps may not be between science and practice or policy communities as these may not even be distinct but lie elsewhere.

One expectation is that there may also be specific organizational forms and institutions that foster closure of gaps between knowledge and practice. For research-based knowledge this may include things like research councils and funding mechanisms for research and development activities, spatial planning policies to encourage co-location of certain combinations of firms or firms with tertiary education or public research institutes. Boundary organizations, for example, may provide a way to regularly bring together the needs and experiences of practitioners with the inventiveness and skepticism of researchers in a mutual relationship that helps create useable knowledge (Guston 2001; Cash et al. 2003). Technology assessments jointly conducted by various actors involved in production and consumptions, including those at risk from harvesting or wastes, are likely to be more effective than expert-only, concerned-community, or industry-driven appraisals. Studies of environmental assessments have underlined the importance of addressing the dynamic tensions between saliency, legitimacy and credibility (Cash 2000; Cash et al. 2003).

Design

This brings us to the main question addressed by this book:

Are some activities, strategies of actors, or institutional arrangements for bringing research-based and other forms knowledge to bear on integrated management of production–consumption systems more effective than others at enabling transformations towards sustainability?

This is an empirical question that could be informed by analysis and comparisons of cases. But it is challenging to tackle head-on so we asked case study contributors a set of smaller, more tractable, questions:

1. What are the main challenges to sustainability? Who are the main actors involved?
2. What are the main sources of knowledge? What knowledge do actors use? How is knowledge shared?
3. What knowledge is most needed for sustainability? If the required knowledge is absent, what could be done to stimulate its creation? Where appropriate knowledge already exists, but is not effectively used what should be done?
4. Which actors are already working to enhance knowledge-action linkages for sustainability? How could their efforts be enhanced and expanded? What are the highest priority actions to narrow gaps between knowledge and action?

This book is made up of a series of case study chapters. In each chapter, the author(s) reflected on these questions for a PCS they knew well extracting what they felt were the key insights about making PCS more sustainable. The case study approach often enabled researchers to engage directly with some of the actors involved in the production, consumption or regulation of specific goods or services and other stakeholders impacted by those processes. This engagement was encouraged as it was hoped it would lead to deeper understanding of what it takes to bring about change.

The case studies in this book include studies focused on developing and developed countries in Europe and Asia. The cases fall into four groups.

The first group deals with lifestyle and livelihood issues (chapters 2, 3 and 4). It is concerned with people at the ends and fringes of PCS, some with too much, and others with not enough. There are three chapters. Michael Maniates writes about "inspired restraint" illustrating with the public campaigns in the USA to reduce consumption and gain back quality time. Uchita de Zoysa's chapter talks about limits to growth and the challenges of consumer activism, in particular, for sustainable consumption in developing countries. The contribution by Fritz Reusswig presents and illustrates a framework for assessing lifestyle changes using case studies of wind energy and carbon footprint labeling.

The second group deals with energy systems (chapters 3, 5 and 6). Two chapters are about experiences in Germany. The contribution by Doris Fuchs and Sylvia Lorek looks carefully at information and knowledge issues and how they limit and influence electricity supply choices of consumers. Part of Fritz Reusswig's chapter explains how the actors and policies involved in promoting wind energy in Germany shifted over time. The third contribution in this section by Rajesh Daniel and colleagues looks at the agrofuel policies of Thailand's government over the last 10 years, exploring the rationales and assessing the practical consequences for sustainability of energy and farming systems.

The third group deals with food systems, one of the better-studied domains in PCS (chapters 7-10). The first by Dao Huay Giap and colleagues analyzes the differences between research and policy in Thailand's shrimp aquaculture industry as it has responded over time to both domestic and international challenges to become more sustainable. The second chapter by Benjamin Nölting looks at how organic food niche markets have developed in eastern Germany with emphasis on what knowledge is being shared and how it is understood by different actors. The third case study by Alga Kalfagianni looks at how the pork consumption chain in Europe has developed and critically analyzes the potential for shifting from traceability schemes primarily concerned with health and safety issues to also incorporating environmental sustainability concerns. The fourth chapter by Arnold Tukker reviews understanding of certification schemes in coffee through the lens of innovation and transition ideas explaining why some niches are unlikely to expand much whereas others could but still struggle to maintain high sustainability criteria.

The final group deals with tourism as an example of PCS that deals in a lot more intangible goods and services, but also has significant social and environmental impacts. There are three chapters (11-13) drawing on experiences of developing country destinations in Malaysia, Thailand and India. The contribution by Giok Ling Ooi describes some of the politics and pitfalls of tourism-site development that fails to maintain natural amenity values. The second chapter by Hannarong Shamshub summarizes a multi-dimensional analysis of the tourism system in Thailand drawing several important PCS conclusions for sustainable tourism policies. The third chapter by Ligia Noronha focuses on tourism development in Goa, noting how scientific knowledge, advocacy and investment politics interact to shape policy and coastal land-use.

The book ends with a synthesis chapter that draws extensively on the key insights from the 12 case studies to draw wider lessons about the challenges and strategies for linking knowledge and action in PCS.

Acknowledgements Many of the ideas in this paper grew out of, in the first instance, a small gathering of scholars and activists in Chiang Mai, Thailand, in October 2004 which was then followed by the creation of the SPACES network and a second meeting of some of the same people in Siegburg, Germany, in October 2005. A broader conference introduced initial findings in Chiang Mai in January 2007. The workshops were sponsored by a grant from the David and Lucille Packard Foundation. Their support is gratefully acknowledged.

References

Barber J (2003) Production, consumption and the world summit on sustainable development. Environ Dev Sustain 5:63–93

Cash D, Clark WC, Alcock F, Dickson NM, Eckley N, Guston DH, Jager J, Mitchell RB (2003) Knowledge systems for sustainable development. PNAS 100:8086–8091

Cash DW (2000) Distributed assessment systems: an emerging paradigm of research, assessment and decision-making for environmental change. Global Environ Change 10:241–244

Cash DW, Borck JC, Patt AG (2006) Countering the loading-dock approach to linking science and decision making. Sci Technol Hum Values 31:465–494

Clark G (2007) Evolution of the global sustainable consumption and production policy and the United Nations Environment Programmes' (UNEP) supporting activities. J Cleaner Prod 15:492–498

Clay JW, Dufey A, MacGregor J (2005) Leverage points for encouraging sustainable commodities. International Institute for Environment and Development, London

Duchin F, Hubacek K (2003) Linking social expenditures to household lifestyles: the social accounting matrix. Futures 35:61–74

Gunderson LH (2000) Ecological resilience – in theory and application. Annu Rev Ecol System 31:425–439

Guston DH (2001) Boundary organizations in environmental policy and science: an introduction. Sci Technol Hum Values 26:399–408

Hamilton C (2003) Growth fetish. Allen and Unwin, Crows Nest, Australia

Hamilton C, Denniss R (2005) Affluenza: when too much is never enough. Allen and Unwin, Crows Nest, Australia

Hertwich E (ed) (2002) Life-cycle approaches to sustainable consumption. International Institute for Applied Systems Analysis, Laxenburg, Austria

Hertwich E (2005) Life cycle approaches to sustainable consumption: a critical review. Environ Sci Technol 39:4673–4684

Hobson K (2003a) Consumption, environmental sustainability and human geography in Australia: a missing research agenda? Aust Geogr Stud 41:148–155

Hobson K (2003b) Thinking habits into action: the role of knowledge and process in questioning household consumption practices. Local Environ 8:95–112

Kates RW (2000) Population and consumption: what we know, what we need to know. Environment 42:10–19

Kates RW, Clark WC (eds) (1999) Our common journey. National Academy, Washington DC

Lebel L (2004) Transitions to sustainability in production-consumption systems. J Ind Ecol 9:1–3

Lebel L, Lorek S (2008) Enabling sustainable production–consumption systems. Annu Rev Environ Resour 33:241–275

Leiserowitz AA, Kates RW, Parris TM (2006) Sustainability values, attitudes, and behaviors: a review of multinational and global trends. Annu Rev Energy Environ 31:413–444

Lorek S, Lucas R (2003) Towards sustainable market strategies: a case study on eco-textiles and green power. Wuppertal Institute, Wuppertal, Germany

Lorek S, Spangenberg JH (2001) Indicators for environmentally sustainable household consumption. Int J Sustain Dev 4:101–119

Manoochehri J (2002) Post-rio "sustainable consumption": establishing coherence and a common platform. Development 45:47–53

Meadows DH (1999) Leverage points: places to intervene in a system. Sustainability Institute, Hartland, VT

Murphy J (2001) From production to consumption: environmental policy in the European Union. In: Cohen MJ, Murphy J (eds) Exploring sustainable consumption: environmental policy and the social sciences. Pergamon, United Kingdom

Myers N, Kent J (2003) New consumers: the influence of affluence on the environment. Proc Natl Acad Sci 100:4963–4668

Narayan D, Walton M, Chambers R (eds) (2000) Crying Out for Change. Oxford University Press, Oxford

Ostrom E (2003) How types of goods and property rights jointly affect collective action. J Theor Polit 15:239–270

Princen T, Maniates M, Conca K (eds) (2002) Confronting consumption. MIT, Cambridge

Spaargaren G (2003) Sustainable consumption: a theoretical and environmental policy perspective. Soc Nat Resour 16:687–701

Spangenberg JH, Lorek S (2002) Environmentally sustainable household consumption: from aggregate environmental pressures to priority fields of action. Ecol Econ 43:127–140

Stern PC, Dietz T, Ruttan VW, Scolow RH, Sweeney JL (eds) (1997) Environmentally significant consumption: research directions. National Academy, Washington DC

Tukker A, Jansen B (2006) Environmental impacts of products: a detailed review of studies. J Ind Ecol 10:159–182

United Nations (UN) (1992) Agenda 21; Results of the World Conference on Environment and Development. United Nations, New York

van Kerkhoff L, Lebel L (2006) Linking knowledge and action for sustainable development. Annu Rev Environ Resour 31:445–477

WSSD (2002) World summit on sustainable development: plan of implementation. United Nations, New York

Chapter 2
Cultivating Consumer Restraint in an Ecologically Full World: The Case of "Take Back Your Time"

Michael F. Maniates

Ours is an ecologically full world of some 6.5 billion people, more than half of whom live in material poverty. It is a world where the capacity of environmental systems to absorb abuse while delivering vital goods and services is more than fully taxed. Current patterns of global consumption deplete natural resources (like fisheries or ground water) faster than they regenerate. Prevailing networks of production create waste in volumes that overwhelm the absorptive capacity of natural systems (leading to problems like climate change). Taken together, 1.4 planets of ecosystem capacity and natural resource stocks are required to sustain human society, and we are quickly heading to 1.5 and beyond.[1] Growing affluence among the world's poor is yielding a class of "new consumers" for whom automobiles, a diet rich in meat, and larger homes with more possessions are the looming norm,[2] and transnational corporations faced with saturated markets in the rich world work diligently to cultivate new consumer appetites among the new consumers of the poor world. Humanity seems locked on a collision course with massive ecological decline, destabilizing crisis, and authoritarian (even draconian) social and political response.[3]

Not, in short, a very pretty picture.

It is not, however, an inevitable picture. Activists and scholars around the world, working alone or in loosely coordinated networks, are building the foundation for environmentally sustainable systems of production and consumption. Their work cuts through treacherous, politicized terrain. Slowing the assault on climate, de-carbonizing the energy economy, transforming agriculture and reviving critical

---

M.F. Maniates
Professor of Political Science and Environmental Science, Allegheny College, USA
email: michael.maniates@allegheny.edu

---

[1] See, for example, the Global Footprint Network <http://www.footprintnetwork.org>

[2] For example Norman Myers and Jennifer Kent, *The New Consumers: The Influence of Affluence on the Environment*, Washington, D.C.: Island Press, 2004.

[3] For a recent treatment of these possibilities, and at least one possible response, see Robert Hopkins, *The Transition Handbook: From Oil Dependency to Local Resilience*, White River Jct., Vermont: Chelsea Green Publishers, 2008.

L. Lebel et al. (eds.), *Sustainable Production Consumption Systems: Knowledge, Engagement and Practice*, DOI 10.1007/978-90-481-3090-0_2,

global fisheries (to name but a few looming challenges) demands more than massive investments in renewable energy infrastructure, or consumer commitment to easy, cost-effective environmental measures like replacing inefficient light bulbs or more aggressive recycling. Nothing less than a new global compact is necessary, one where the overconsumers of the world deliver significant reductions in resource throughput and material accumulation, this in order to create "ecological space" for increasing consumption by the world's poor – and where, in turn, the global underconsumers explore development paths of low-consumption, high-prosperity living.[4] This is contraction and convergence on a grand scale: Contraction of the consumption by the rich as the foundation for the convergence of consumption levels by all at some sustainable level.

At first blush, *any* talk of contraction and convergence seems hopelessly naïve. ("You'll never get the rich to cut back," is one reflexive response; "the poor will never show restraint" is another; "contraction and convergence requires massive value change or some deep, mobilizing crisis" and "Americans will never sacrifice without a crisis" are other common reactions.) It's no wonder that most people who work on issues of sustainable consumption and production shy away from the question of "how much is enough." Where, after all, are the potent research questions – those that generate grants, drive publications, or influence policy – if the desire for ever-escalating consumption is hard-wired in the human psyche or part of deeply held value sets? Who aspires to research and activism that is intrinsically coercive, or that would promote policies of reduced consumption that fly in the face of human desire? Better, many conclude, to focus on "realistic" and tangible responses to ecological overshoot, such as the development of new production technologies capable of accommodating escalating consumption and lower environmental cost, or economic instruments that might shift consumption toward more environmentally benign products, or education and public-information projects that might, over time, reshape values. And, indeed, this is the bulk of the work now occurring under the flag of "sustainable consumption."

What appears to be idealistic or naïve is, alas, coldly realistic. For without significant contraction in material consumption among the rich and inspired restraint by the poor, other important measures to lighten human's footprint on the planet – from individual efforts to live more environmentally to ambitious infrastructure programs to reshape energy systems – will be swamped by the juggernaut of ever-escalating global consumption. Coercion and sacrifice could then become the norm as the ability of ecosystems to provide critical goods and services

[4] A rich literature exists on "contraction and convergence," and on the limits of technological change and consumption shifting alone to fully blunt the dynamics of ecologic overshoot. See, for example, Guha, *How Much Should a Person Consume?* Berkeley: The University of California Press, 2006; Myers and Kent, *op.cit.*; Princen, Maniates, and Conca, *Confronting Consumption*, Cambridge, MA: MIT Press, 2002; and Wolfgang Sachs and Tilman Santarius et al., *Fair Future and Limited Resources*, London: Zed Books, 2007. The nearby figure on "contraction and convergence" is taken from Sachs et al.

decline.[5] New technologies, consumption shifting and education: they all blunt the edge of growing global consumption and buy time to address the deeper question of how to challenge a spreading global culture of consumerism. But they are not sufficiently potent to align consumption with global ecosystem capacity to process waste, regenerate renewable resource systems, and support human economic activity. In the face of exponentially growing consumption in a world beset by numerous environmental threats to human well being, nothing less radical and realistic than informed, careful struggle toward a global norm of consumer restraint will prove sufficient.

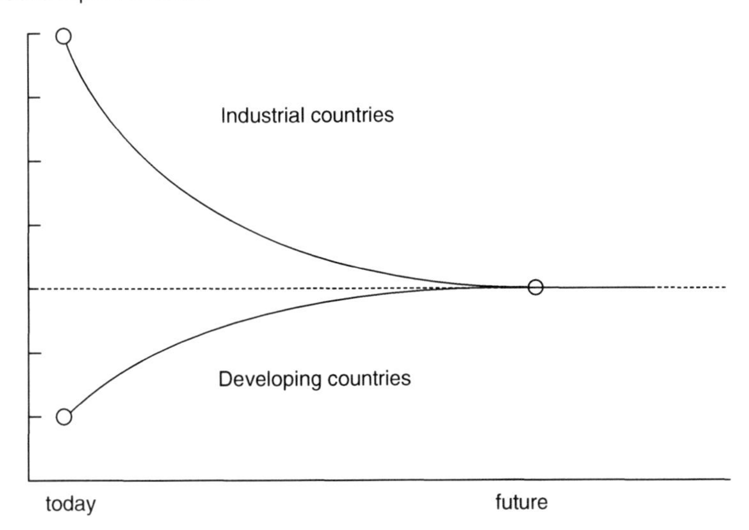

The barriers to such norm-building are numerous and powerful – which is another way of saying that the field is rich with provocative research questions. The path must be paved by vigorous discussion and debate: sustained, global, dynamic, and animated by innovative local, national, and transnational initiatives that explore avenues for greater prosperity through lower levels of consumption.[6] Perhaps the greatest barrier to these explorations is the hegemonic view that *any* decline in

[5] In stark contrast to contemporary environmental discourses, which often view "crisis" as a necessary, even desirable precursor to policies of environmental sustainability and social justice, earlier environmental writers highlighted the undesirable consequences of crisis: authoritarianism, anti-democratic impulses, the search for quick fixes, even those that undermine essential liberties or civil rights. Today's scholars of environmental sustainability could benefit from a close reading of this earlier literature. See, for example, Robert Heilbroner, *An Inquiry Into The Human Prospect: Updated and Reconsidered for the Nineteen Nineties*, New York: R.S. Means, 1991 (originally published in 1980), William Ophuls, "The Scarcity Society," *Harper's Magazine*, April 1974, pp. 47–52, and Ophuls, *Ecology and the Politics of Scarcity*, New York: W.H. Freeman, 1977.

[6] Guha, *op. cit.*, offers a template for how to think about such a process in his discussion of escalating consumption in India.

individual or household consumption *always* leads to a loss of happiness, or to sacrifice. That this article of faith is so easily received, even as a growing body of transnational research demonstrates the contrary, frustrates those who'd hope for research-driven changes in public policy and popular perception. From the simple standpoint of individual human happiness and socially collective well-being, more isn't always better. Less is sometimes best. Yet, if "material restraint" or "reduction in consumption" remains synonymous with "sacrifice and pain," no policymaker or politician will rise to the challenge of forging new avenues for debate and change. Policymaking on the consumption question will come only in the face of intractable crisis, long after windows of creative, effective, and anticipatory consumption policy have closed.

To what extent can determined activism bolstered by strategic research undermine the view that happiness is linked to ever escalating consumption? How might public policy and new institutions that offer individuals and communities opportunities to consume less in ways that enhance immediate happiness and overall life satisfaction best be identified, and then injected squarely in the midst of public conversation? Where do the pressure points for a shift to sustainable consumption lie in a politics of the global north that celebrates consumption? And how, for the purposes of this volume, might additional research facilitate meaningful political change in support of an agenda of contraction and convergence? This chapter touches on these questions by exploring "Take Back Your Time" (TBYT), a public-policy initiative now underway in the United States that aspires to build a participatory politics of consumption reduction. Built around the notion of "time famine," TBYT argues that politically constrained choices around work and leisure in the United States make it especially difficult for United States (US) consumers to exercise restraint in their consumption choices. If offered alternate choices, especially choices regarding the structure of work, Americans would consume less *in the rational pursuit of their own happiness*. Even modest success of TBYT's agenda would be an important step in a politics of contraction and convergence that rejects a discourse of sacrifice and deprivation.

Three sections follow. The first orients the reader to the evolution and aims of TBYT. The second describes a recent, revealing shift in the program that highlights several research opportunities (or "knowledge gaps," in the language of this volume) for scholars and practitioners of sustainable consumption. The elements of an alternate research framework around sustainable consumption–production systems emerge in a final section. Each section draws on extended conversation with those within TBYT, others in the US environmental community working on issues of sustainable consumption, and the author's own participation in TBYT.

These sections together offer a story about potential – the potential for activists to challenge the view that more consumption yields greater happiness, and that of researchers to drive an innovative politics of sustainability through targeted work in support of activist agendas. It is a story of pregnant possibility rather than tangible accomplishment, one of shifting plot lines shaped by global financial instability, the rising cost of oil, and the pressure on production–consumption systems dependent upon cheap energy to evolve and adapt. It is, ultimately, a story meant to raise questions.

How, for example, might research expertise best be applied to the task of creating institutions and incentives that join increased happiness with reduced material consumption? What are the payoffs of close collaboration between scholars and practitioners? Perhaps most important, in what ways must "sustainable consumption" as a research enterprise be reimagined if new knowledge is to make a difference fast enough to count?

Making a "Movement" Real

From VSM to TBYT

Anti-consumerist notions of simplicity, frugality, and living with less are nothing new to the US cultural landscape. For the past 200 years, periodic bursts of public interest in frugality and simple living – as both a moral choice and political act – have been followed, more often than not, by even more intense periods of accelerated acquisitiveness.[7] In contemporary US culture, these counter-cultural impulses to live with less cohere in what many call "the voluntary simplicity movement" (VSM). The outlines of today's VSM emerged as part of the "appropriate technology movement" and "back to the land" impulses of the late 1960s and early 1970s (themselves a response to the politically turbulent 1960s), and were best captured by Duane Elgin's 1981 book *Voluntary Simplicity*.

Multifaceted and uncoordinated, VSM became, in the 1980s and 1990s, the banner for a wide range of loosely connected individual actions centered on consumption choices. Action ranged from enacting the advice from newsletters like the *Tightwad Gazette* (tips on how to live life on the cheap) to expressly political objections to structural poverty (even today one can see in the US the vintage 1980s bumper sticker "Live Simply So That Others May Simply Live"). The intensity of these actions varied, from limited consumption shifts among many to radical lifestyle changes by a few, but they typically shared a preoccupation with "taking control" of one's life and escaping, if for just a bit, the demeaning, numbing effects of everyday commercial life.

The rise of the VSM during this time was surely connected to the tightening squeeze on the US middle class. The decline of the US labor movement in the 1980s, the growing number of two-wage-earner households during this period (in response to a "middle-class squeeze" of flattening salaries and rising costs, and changing norms about women in the workplace), and an upward creep in the length of the average work week meant that Americans were spending more time at work even as the proportion of two-income families was growing.[8] As work drew time

[7] David Shi, *The Simple Life: Plain Living and High Thinking in American Culture*, Athens, Georgia: The University of Georgia Press, 2001.

[8] Elizabeth Warren and Amelia Tyagi, *The Two-Income Trap: Why Middle-Class Mothers & Fathers Are Going Broke*, New York: Basic Books, 2003.

and energy away from home and leisure activities, families increasingly characterized their lives as harried and "out of control." At the same time, writers and speakers with something to say about the hidden costs of consumerism and ways of escaping "the rat race" enjoyed increasing visibility. Joe Dominguez and Vicki Robin, for example, became known for their self-help seminars in the late 1980s that led to their 1992 book *Your Money or Your Life: Transforming Your Relationship with Money and Achieving Financial Independence*. John de Graaf's 1997 film *Affluenza* (which documents the hidden costs of consumerism to individuals and families across the US) was, for a great stretch of time, one of the best selling, most watched films shown on the US Public Broadcasting System.[9] Former community-college administrator Cecile Andrews was catapulted to prominence as the author of *The Circle of Simplicity* (1998), a guide to forming neighbourhood study and action groups for removing the clutter of materialism from one's daily life. Robin, Andrews, and de Graaf, together with other writers, filmmakers, and academics, could frequently be seen together at "simplicity meetings" and workshops, sometimes drawing crowds in the hundreds when only a few dozen were expected.

These simplifiers weren't radical, and they weren't fringe. During this period some 15–25% of US households were intentionally "downshifting" by consciously living below their means, refusing to take higher paid work to avoid job stress and long work hours, or voluntarily reducing work hours at an existing job, even in the face of diminished opportunities for advancement and salary increases. A majority of these simplifiers were more highly educated than the average citizen; college degrees are common to this group, and advanced degrees or special skill sets are not rare. Most simplifiers, moreover, were at or near median household income levels before their tilt towards frugality – few, in other words, were "simplifying" by trading their Jaguar in for a BMW. And most, it bears repeating, were motivated by workplace overload or burnout.[10]

The voluntary simplicity movement wasn't really, then, a movement per se. It was instead a vivid example of a distinctly American inclination toward individualized, consumer-centered responses to social ills better addressed collectively, though political action and public debate. There were no formal leaders, no designated spokeswomen and men, no central agenda, and hence no capacity for the kind of focused social action that distinguishes effective social movements. It was politically incoherent, a movement in waiting, one laden with possibility.

A shift came in 2001, when the Fetzer Institute[11] invited 24 "leaders of the simplicity movement" to its conference center to explore how the VSM could be

[9] According to Bullfrog Films, the film's distributor.

[10] For more data and sources, see Michael Maniates, "In Search of Consumptive Resistance: The Voluntary Simplicity Movement," in Thomas Princen, Michael Maniates, and Ken Conca, eds., *Confronting Consumption*, Cambridge, MA: MIT Press, 2002.

[11] "The Fetzer Institute's mission, to foster awareness of the power of love and forgiveness in the emerging global community, rests on its conviction that efforts to address the world's critical issues must go beyond political, social, and economic strategies to their psychological and spiritual roots." From http://www.fetzer.org/

sharpened into a genuine political force. This inaugural meeting was followed by two strategy sessions in 2002 – one in March in Kalamazoo, Michigan, the other in August at Oberlin College in Ohio – that saw the Simplicity Forum grow to 68 members, including several academic researchers (including this author).[12] Though far ranging, the discussions usually returned to one question: how can the millions of people who consider themselves part of the VSM be mobilized on behalf of meaningful policy change? One slowly emerging answer was the "Take Back Your Time" project. John de Graaf, a documentary filmmaker (with the aforementioned "Affluenza" among the productions in his portfolio) living in Seattle, Washington, assumed primary responsibility for the project.

In the years since its launch, TBYT has established an identity and agenda separate from The Simplicity Forum (though a perusal of the websites of both programs shows a close working relationship). de Graaf, working with others, has generated a level of public attention and debate about work, consumption, and sustainability well out of proportion to the meager resources at his command. His is the only organization that consistently addresses the connections among work, leisure, consumption, social capital, and environmental sustainability. Groups ranging from conservative businessmen to college students warm to his message that overwork, lack of vacation, and few if any options for part-time work for part-time pay (while keeping one's health benefits) drive levels of consumption that are unhealthy for people and the environment. Visitors to TBYT's website (www.timeday.org) and its sister project (www.right2vacation.org) quickly grasp the message that working less and consuming less can be a realistic means to a happier life.

And yet, despite a striking level of publicity and conversation, TBYT has generated little actual policy action since its inception. de Graaf, and other environmental leaders interviewed for this chapter, believe that this could quickly shift. Americans are poised for a change, they believe, and ever-escalating levels of consumption are not the only path to individual and national prosperity. The right combination of factors could elevate TBYT's agenda to national prominence.

The TBYT Agenda

TBYT seeks to mobilize political resources in support of state and federal legislation that would allow, and perhaps compel, Americans to trade consumption for leisure. At its birth, TBYT was imagined as a vehicle for tapping middle-class frustration in support of national legislation for an optional 32 h/4 day work week. (Under this legislation, workers would have the right, with some exceptions, to work three-fourth or four-fifth time for three-fourth or four-fifth pay, while keeping the core of employer-provided benefits.) This policy agenda, for reasons described below, failed to generate a critical mass of public support and political action.

[12] Meeting summaries are available from the Simplicity Forum at http://www.simplicityforum.org/congressreports.html

TBYT's policy agenda was subsequently expanded and reordered: Part- and flex-time work (still the core agenda item) was dropped to the bottom of an expanding list of initiatives, and other time-focused initiatives were brought forward. These included mandatory paid family leave, mandatory paid sick-leave for all employees, and mandatory paid vacation. It was hoped that progress on one or more of these initiatives would build momentum for revisiting the 32 h work week, on the assumption that many Americans, if offered the opportunity, would choose less work (and thus reduced consumption) for more leisure time.[13] Since the beginning of the latest recession, TBYT has also been seen again by reporters and activists as a resource for ideas about saving jobs by shortening the work week and spreading jobs around.

The root problem, in TBYT's view, is one of recent history and political economy. Despite warnings after WWII that productivity gains in the US would result in a

[13] TBYT's "Time to Care Public Policy Agenda" is available at www.timeday.org

glut of leisure time (a distinct worry among sociologists of the time), specific political–economic processes drove the US in a different direction: increased productivity gains per worker were translated into the exponential growth of per capita material acquisition with no increase in leisure time, and with little measurable increase in human happiness. This process intensified in the 1980s and 1990s, when a work and spend treadmill[14] energized by easy credit and stagnant wages fueled a mutually reinforcing dynamic of more hours at work, growing compensatory consumption outside of the job, increased debt, and increased work. The interplay of overwork and rising consumption leads to increased levels of stress, undermines family and neighbourhood bonds, generates a rapid increase in "convenience consumption" (e.g. fast food), TV consumption and binge vacationing. The result is escalating environmental degradation[15] and withdrawal from civic life.

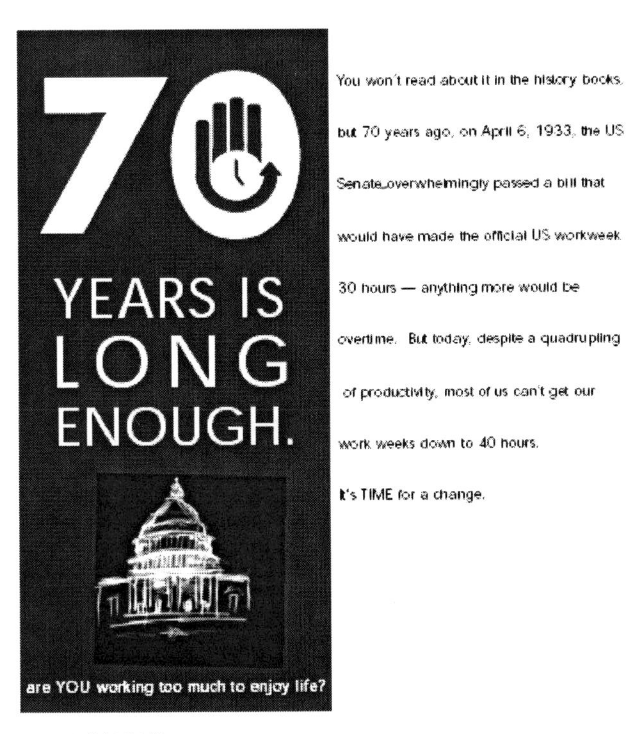

[14] Schor provides an accessible analysis of this treadmill. See Juliet Schor, *The Overspent American: Why We Want What We Don't Need*, New York: Harper, 1999.

[15] See, for example, "Are Shorter Work Hours Good for the Environment?," at http://www.cepr.net/documents/publications/energy_2006_12.pdf

From TBYT's perspective, many Americans grasp this dynamic and yearn for a remedy. They work, as individual consumers and workers, toward piecemeal solutions. Yet their stubbornly individualistic efforts to change this situation yield little of lasting consequence absent broader institutional change. For example, a few employees at a firm might negotiate flexible or part-time work hours (drawing on an array of resources on effective strategies for "getting your way"[16]), but their success can make it difficult for other employees to later cut similar deals of their own – and indeed, such arrangements could exacerbate the overwork of others. Even when employers are accommodating, the available choices are limited. Because US employment is comparatively "blocky" – one is either unemployed or works 46 h/week – and because the best health benefits come with full-time employment, it is difficult for people eager to trade reductions in consumption for less time at work to pursue this option.

[16]For example Lynn Berger, *The Savvy Part-Time Professional: How to Land, Create, or Negotiate the Part-time Job of Your Dreams*, Sterling, VA: Capital Books, 2006, or Page Hobey, "When No Could Mean Yes: How to Negotiate a Flexible Work Schedule," at http://www.mommytrackd.com/no-means-yes

TBYT isn't so naïve as to believe that all Americans share its perspective on the costs and perils of overwork. It maintains, though, that mainstream politics in the United States has for too long attributed overwork – long hours, little vacation, restricted family and sick leave – to something virtuous and important: a valued cultural holdover from the days of the frontier, perhaps, or evidence of a uniquely "can do" American approach to life. In reality, the drivers, costs, and possibilities are more nuanced. Many Americans feel trapped by too much work and too few choices in the workplace. Many engage in "downshifting" and speak to a desire to have more time outside of work, even if it means making less.[17] Given the choice, many such individuals would choose to work less, earn less, and have more leisure time for broadly non-material activities. These early innovators would serve as a model for others – as living examples of any number of lower consumption/higher prosperity ways of living – and help drive deeper, more animated debate about non-material sources of satisfaction and prosperity.

How, though, might this slice of America best be mobilized by a new activist group with limited resources? For TBYT, the answer came straight out of the mainstream literature in environmental policymaking: create a focusing event, a big splash, a moment that crystallizes public concern and focuses the attention of political elites.[18] Inspired by the cultural impact and political success of the first "Earth Day" in 1970, de Graaf and others in the Simplicity Forum hit upon the notion of "Take Back Your Time Day." Observed on October 24 since 2003, "TBYT Day" commemorates what would be the final day of work for the year if Americans enjoyed, on average, the same vacation, sick leave, and holiday benefits of their European counterparts.

[17] For example Schor, *op cit.*, and Maniates, *op cit.*

[18] The notion that critical "focusing events" – a deadly smog, rivers aflame, massive oil spills, energy price spikes – drive progressive environmental policy runs through much of the sustainable consumption literature. One consequence is the perception that policy progress requires crisis, which can be a debilitating assumption when the crises at hand are potentially catastrophic and irreversible. Another consequence of the focusing event logic is the assumption that viable political pressure can be mobilized by *creating* a focusing event (Greenpeace with its photogenic Zodiac speedboats doing battle with lumbering whaling trawlers is the classic example). Lost in the ensuing discussion about whether to manufacture focusing events or wait for the catastrophes on the horizon is a critical assessment of the "focusing event" logic itself. Do dramatic events indeed lie at the root of striking changes in environmental (and other) policy? Or are other, more pivotal forces, in play? For a defining articulation of focusing-event politics regarding environmental policy, see Anthony Downs, "Up and Down with Ecology: The Issue Attention Cycle," *The Public Interest*, Volume 28 (Summer 1972), pp. 38–50. For a recent questioning of this focusing-event logic, and the price environmental scholars pay for so readily accepting it, see Michael Shellenberger and Ted Nordhaus, *Break Through: From the Death of Environmentalism to the Politics of Possibility*, New York: Houghton Mifflin, 2007.

Buzz but No Bounce

From a publicity/public awareness standpoint, Take Back Your Time Day has met de Graaf's expectations. Dozens of colleges and universities have adopted the Time Day project as a teaching tool, more than 150 communities have initiated Time Day observances, and near a 1,000 articles about time famine have appeared in major newspapers and other media outlets, especially around October 24th – all this in a national political environment generally hostile to debate about the work and worker's rights. The Time Day website – well crafted, frequently updated, visually compelling – receives thousands of hits a month. *Take Back Your Time: Fighting Overwork and Time Poverty in America*, a 2003 resource book edited by de Graaf and touted as "The Official Guide to Taking Back Your Time," continues to sell many years after its publication. The same is true for *Affluenza*, a trade book that captures many of the themes of de Graaf's two films for the Public Broadcasting System: "Affluenza" and "Escape from Affluenza." For an activist group of its size

(essentially de Graaf – employed part-time as Executive Director – and a few volunteers), the sustained "buzz" generated by Take Back Your Time Day would be the envy of much larger, better financed activist groups.

One source of this growing attention was TBYT's expanding public policy agenda, which slowly grew to include promotion of more sick leave, better vacation policy, release from mandatory overtime, and even Election Day as a federal holiday. Simultaneously, the range of suggested activities for TBYT Day exploded, making necessary an extensive web page at timeday.org to list all the possibilities.[19] With multiple activities and messages, TBYT spoke to more constituencies. "We've clearly struck a nerve with Take Back Your Time Day," says Gretchen Burger, then a national staff person for TBYT, in a 2006 phone conversation. "I'm amazed by the number of emails and phone calls that I receive, from people all over the country, wanting to learn more or become part of this initiative."

And yet, in seeking to be many things to many people, TBYT sacrificed much of its ability to mobilize people around a single, clear and distinctive problem and set of actions. The diffusion of message and loss of focus wasn't planned. Rather, it arose incrementally as the organization sought new opportunities to broaden its base and cultivate funding. Financial support, in fact, has been a long-standing problem, one that de Graaf believes prevents TBYT from establishing itself as a national policy force. At fault, he suggests, are the ideological norms of major foundations and the framing of their philanthropic initiatives. Conservative foundations, says de Graaf, appreciate the TBYT focus on family time and the cultivation of civic virtue in place of rampant consumerism, but they balk at the implied regulation of workplace hours. Left-leaning foundations understand the connection among time famine, environmental degradation, and overconsumption, but don't know how to fit funding for the project into a template that separates these connections into discrete program silos. Interestingly, other activists and funders interviewed for this study (some working for sustainable-consumption NGOs, others more tightly linked to the foundation community), disagree. They believe that ample funding could be available to TBYT if it focused its message, developed a political strategy for mobilizing support around a legislative agenda (a strategy that moves beyond "celebrating" TBYT Day), and focused its attention on a handful of core constituencies.

Upon reflection, de Graaf is inclined to agree. "Perhaps we've tried to be all things to all people," he said in early 2007, "and, as a result, we haven't found one or two items around which a critical mass of people will rally." But generating a potent strategy and working closely with dedicated constituencies takes staff, and TBYT is a distinctly two- , sometimes three-person operation at best. It hasn't had the capacity to capitalize, in politically tangible ways, on the enthusiasm generated by scores of TBYT events that regularly occur around the United States. And yet, other political movements have gotten off the ground with less coordinated support and publicity. The "Student Anti-Sweatshop Movement" in the United States, which swept university campuses in 2002 with an agenda of ridding bookstores and

[19] See http://www.timeday.org/tbyt_day.asp

university clothing shops of products made under sub-standard working conditions or sweatshop wage rates, is but one example. That activist endeavor had a tight, simple message (no sweatshop products will be sold in college bookstores) and sought to mobilize a relatively few number on people (students at just a few campuses) around very specific activities.

Examples like these weren't lost on de Graaf and those around him.

Their answer: Take a Vacation.

From "Politely Taking" to "Demanding Rights:" Moving from Mass Appeal to Targeted Action

Despite its considerable success in focusing public attention on time famine and overconsumption, TBYT suffered from two problems, one of practical politics, the other of political psychology. Practically speaking, the media-savvy choice to center the initiative on an annual "Take Back Your Time" Day left participants unsure about how they might best foster meaningful change. TBYT's range of suggested actions were largely polite and symbolic, and usually too broad and often too trivial to create a sense of solidarity around the felt possibility of meaningful change. The political psychology of the initiative was equally ambiguous. Should participants be forceful, even angry, about the erosion of leisure time and the loss of worker control, and seek ways to ameliorate the situation? Was TBYT essentially a movement of resistance and protest? Had something been taken from us (our time), which we now had the right to reclaim? Or was the program more of a lighthearted affair meant to foster conversational openings about the benefits of better work–life balance?

TBYT's experience underscores the difficulty of mounting any effective political action in the absence of a clear sense of rights – rights enjoyed, revoked, violated, or withheld.[20] Yes, TBYT flirted with the sense of rights in its discussion of US labor history, pointing out that a 30 h work week was once on the cusp of passing as the law of the land. But de Graaf and others were never comfortable asserting that right to a reduced work week. Even the right to the option of working 32 h a week was deemed too risky at a time of insecure affluence among the middle class. TBYT was left, then, to float a policy agenda of guaranteed leave for birth or adoption, guaranteed sick leave, and guaranteed vacation, a day off for national elections, and making it easier to choose quality part-time work. The supporting language was timid ("Let's bring the United States up to the standards already in place in all other industrialized countries") and inevitably scattershot.

[20] James Jasper, *The Art of Moral Protest: Culture, Biography, and Creativity in Social Movements*, Chicago, IL: The University of Chicago Press, 1997.

By the spring of 2008, the limitations of a rights-averse strategy were fully apparent. Joining forces with Joe Robinson (author and director of a small NGO called "Work to Live"), de Graaf launched "Right2Vacation.org," a sister initiative of TBYT. The new website launched with a national poll showing broad public support for mandatory, annual, paid vacations in the United States, a marked departure from current conditions.[21] The shift in political tone was palpable. Rather than politely suggest that workers reassert some claim over their time in any number of small ways, TBYT is urging Americans to assert their *right* to vacation time, a right others readily enjoy but which has been denied to US citizens. The tagline "The United States remains the only industrial country without a law guaranteeing paid vacations" permeates their public outreach materials. It's supported by a four-pronged argument: Loss of vacations (1) raises health-care costs, (2) reduces productivity, (3) undermine desirable social outcomes (time with family, connection to neighbours, civic engagement) and (4) costs the travel and vacation industry, with considerable loss of public engagement with nature. Each argument speaks to different constituencies but the focus rests on a single action: recover the right recognized in *all* other industrialized countries for people to enjoy time for relaxation and renewal.

[21] Paid (or, indeed, even unpaid) vacations are not mandated in the United States, and only 14% of working American take 2 weeks or more of vacation a year.

Vacation?

The notion of "vacation" as the springboard for a more aggressive politics of leisure and reduced material consumption may seem odd. After all, aren't vacations resource-intensive affairs marked by long airplane flights and too many trips to the buffet? Aren't they part of the problem, not the solution?

Not necessarily. First, the binge vacationing common to the United States – 3 or 4 day trips to exotic locales, marked by intense consumption – is as much a reaction to the paucity of vacation time and mounting job stress as anything else. Vacationing needn't be hurried and hedonistic. If workers knew they had an extended paid period of time away from work, they'd be less inclined to concentrate their recreation into short bursts of activity that can leave one feeling less refreshed at the end of it all.

Second, paid vacation doesn't mean more income; one's ability to consume wouldn't expand from paid leave (unless one took a second job during this vacation period). Indeed, if the productivity gains of paid vacations failed to offset the costs of these supported leaves, American workers might find themselves paying for their vacation leave through more modest increases in salary. The income effect of a mandatory vacation plan is not openly discussed in TBYT and Right2vacation materials. The operating assumption, however, is that employers would be willing to absorb the costs associated with this plan, as long as any such costs are fully defrayed by the productivity gains (rested workers produce more per hour) and reduction in health-care costs (less job stress means lower health care needs) that come with increased vacation.

TBYT's muscular discussion about rights will surely alienate some of its past supporters. It opens the door, though, to collaboration with US businesses that currently provide paid vacation and life–work balance programs (often to enhance productivity, reduce health-care costs, and stem worker turnover). Should these programs became mandatory, those businesses that already provide them would face no transition costs, and might therefore enjoy a competitive advantage (at least in the short-term) over their less enlightened competitors. Wouldn't these businesses thus find it in their interest to support "the right to vacation?" That's de Graaf's hope.

Forceful conversation about the injustice of limited vacation could also be a catalyst for action on college and university campuses. de Graaf has visited scores of campuses over the past several years; he's an engaging speaker who generates considerable enthusiasm. Yet he has little to show for it in terms of sustained student activism. A narrower, more compelling focus on vacation could be the game-changer, especially in light of reports that today's students value work flexibility and free time far more than their predecessors.[22] Engaging students in a concrete struggle for workplace reform could drive change across the range of TBYT's public-policy agenda.

Finally, tough talk about vacation invites novel coalitions with the travel industry and major environmental groups. The decline in visitation to US forests and national parks, as fewer Americans engage in camping and hiking, offers one example.[23] The drivers of this decline are unclear, though de Graaf and others believe that one culprit is the increased fragmentation of vacation time: a day here and a day there, rather than two or three solid weeks away from work, makes it hard to load the family into the car and head for the wilds. Environmental groups working on public lands and wilderness issues are concerned, since reduced visitation could translate into declining public support for protecting natural lands. It's no accident, then, that Sierra Club Productions, in mid-2008, chose to support the making of a documentary (by de Graaf) on the importance of vacations to the national life. (Contrast this support against the general reluctance of environmental groups to fund or co-sponsor the earlier programs of TBYT.) To the extent that more vacation time could mean more travel – and perhaps more hiking, camping, and lower-cost eco-travel (for, recall, that additional vacation time doesn't come

[22] See, for example, P. Trunk, "What Gen Y Really Wants," *Time Magazine*, July 5, 2007, at http://www.time.com/time/magazine/article/0,9171,1640395,00.html

[23] See, for example, MSNBC's report on "Visitors to National Forests on the Decline," November 29, 2008. <http://www.msnbc.msn.com/id/27970449/>

with additional income) – a potent coalition among travel specialists, environmental groups, outdoor equipment businesses, and nature educators is waiting to be born. The birthing event could prove to be the "'Vacation Matters' National Summit," held in August 2009 in Seattle.[24]

Each of these possibilities – corporate collaboration, environment–travel coalitions, and student engagement – present a set of "knowledge gaps" with which TBYT/Right2Vacation is now struggling.

One Research Need: Corporate Ecology

State regulation over economic life benefits some while hurting others. Rules that limit a particular kind of pollutant, for example, will penalize those industries (and its stockholders) whose production facilities generate the bulk of the offending compound. But this same set of rules will be a boon for businesses that manufacture pollution-control equipment or monitoring devices. Getting new policy in place depends upon understanding the complex "ecology" of winners and losers in the business world, on identifying the winners, and on marshaling their support at critical points in the policy process.

TBYT is no stranger to these realities of regulatory politics. Indeed, over the years it has been in conversation with several corporate entities (e.g. Perrier, Sam Adams beer, Blockbuster Video, and Beringer Wines) that would stand to gain from workplace regulation that created more leisure time. For a time, Panera Restaurants (a self-serve coffee-house and sandwich chain) was a corporate sponsor of TBYT, and Beringer Wines highlighted TBYT on its corporate website. Both stood to benefit from a "slower" lifestyle, one where workers had more time for conversation and relaxation. Beringer's motivation, as reported on the advertising analysis website "AdRant," was straightforward:

> Wine maker Beringer recently conducted a study that found more than half of respondents don't leave work on time but 28% would if they had social engagements with friends and family planned. With that nugget of information, Beringer has launched "Living 5 to 9," a nice play on the 9 to 5 grind and a new website which aims to help people manage their time better. Oh, and let's not forget the marketing angle here. The more Beringer helps people manage their time, the more likely they will leave work on time thereby allowing them time to stop at the store on the way home and buy more Beringer products. Everybody wins with this one.

Getting people to leave work early so that they can drink wine with friends may seem to trivialize TBYT's policy agenda. Still, Beringer's campaign and its connection to TBYT (which they initiated, not TBYT) is suggestive of the corporate alliances that could emerge around the current thrust toward "vacation rights."

[24]See "Let's Create A 'Vacation Matters' National Conference for 2009" at<http://timeday.org/right2vacation/vacation.asp>

Discerning the outlines of an effective corporate coalition calls for three dimensions of knowledge generation and/or procurement:

- *Identification* In the US political economy, which corporate actors (domestic and transnational) up and down critical commodity and knowledge chains stand to gain the most from a political movement that champions reduced overall consumption in service of increased leisure time? What important analytic and strategic categories emerge (e.g. market share, regional influence, political power, culture of innovativeness) when assessing these potential corporate players? And which nascent industries and economic enterprises that might not yet figure on a list of existing corporations might be catapulted into economic prominence by virtue of a successful "Right to Vacation" agenda?

- *Coalition Articulation* What does theory, but more likely case-study analysis, say about how to best build coalitions of corporate actors in service of a public-policy agenda in support, first, of more paid vacation and, later, greater flexibility in workplace choices? Are business entities at some points in commodity chains more likely to productively participate in coalitions than others? Since many of the corporations involved will be transnational, what existing coalitions might serve or be modified to advance the TBYT agenda? Is there something special about a low-consumption/high-leisure agenda that argues for uncommon approaches to corporate organizing?
- *Inoculation* Finally, how have existing non-governmental organizations with significant corporate sponsorship maintained intellectual and programmatic

autonomy? How, in other words, can an initiative like TBYT that would organize key elements of the corporate sector avoid the appearance or reality of being co-opted?

de Graaf acknowledges the critical importance of such research and would embrace it. He and others working with him know of no systematic research that identifies those elements of the business sector that would be sympathetic to lower consumption/higher prosperity trajectories, much less research that suggests effective ways of organizing a coalition. Knowledge generation in this area could yield significant political payoff.

Another Research Challenge: Catalyzing Student Networks

In an undergraduate research project[25] recently completed for TBYT, Jaclyn Stallard of Allegheny College draws on social-movement theory and a detailed case study of the "Student Anti-Sweatshop Movement"[26] to suggest ways in which TBYT might mobilize enduring student support across the country for its time–famine initiatives. Stallard's work identifies strong resonance of many of TBYT's policy concerns with the interests (and fears) of undergraduate college/university students. Chief among these is the conviction among students that the flexible work schedule they enjoy in school is something they'd struggle to create in the work world. The freedom to trade work (and thus income, and consumptive capacity) for leisure time emerged a consistently attractive option. Stallard's research reveals, however, that TBYT's earlier public-policy agenda was too diffuse to mobilize students. Necessary, in Stallard's view, is a single focus with a strong moral component that connected to students' daily lives. Stallard points to many examples where student activism around an issue spread like wildfire across campuses, and argues that something like the "right to vacation" could have similar effect.

Stallard's work points to three especially important areas for future research:

- *Issue Resonance and Framing* Among university students in both the global north and south, what are the key attitudes and concerns with respect to consumption, overconsumption, job satisfaction, lifestyle, and leisure time? With this knowledge in place, how can initiatives like Right2Vacation and TBYT be narrowed, highlighted, or recast to connect with student sensibilities?

---

[25] *Unlocking the Origins of Political Opportunity: Student Roles and Power Potential in Social and Environmental Movements*, Senior Research Project, Dept. of Environmental Science, Allegheny College. An abstract is available at http://webpub.allegheny.edu/dept/envisci/ESInfo/comps/2006abstracts.html#stallard Copies of the larger report are available upon request from Michael Maniates.

[26] See Peter Dreier and Richard Appelbaum, "The Campus Anti-Sweatshop Movement," *The American Prospect*, November 30, 2002, and the homepage of the US Student Anti-Sweatshop Movement at http://www.studentsagainstsweatshops.org/

- *Network Building* Student activist networks, either US centered or transnational, have been remarkably effective in elevating the political salience of public-policy issues and marshalling political pressure at key moments. Stallard's work identified the Student Anti-Sweatshop initiative as a model, both for its rapid growth and success, and for its transnational character as well. The "anti Coca-Cola" network is an example of another transnational network that took the university student community by storm – but unlike the Anti-Sweatshop movement, this network originated in the global south.[27] What are the key elements of especially successful networks of student activism? How congruent are these elements with the capacity of TBYT and similar organizations to foster student activism? What are the organizational and financial costs to launch a network, with a reasonable hope for policy success?
- *Connecting to Strategic Planning* Higher education in the United States faces a daunting demographic reality over the next two decades. Nationally, the number of students seeking admission to colleges and universities will peak in 2009–2010, level off for a handful of years, then decline – slowly at first, and then with increasing pace. Most colleges and universities, especially those with limited endowments or constrained markets, are scrambling to develop strategic plans and marketing niches that will allow them to thrive amidst these new demographic realities. Many are looking to frame their historical mission – education for the social good, training new workers, generating research products – around new opportunities and realities. The need to do so has only been intensified by the global economic crisis. Which institutions of higher learning would be most amenable to incorporating ideas about work, leisure, stress and policy change in its strategic planning? How might colleges and universities seeking special relevance to emerging economic conditions be identified and introduced to TBYT's agenda? How, in particular, might the nation's ~ 500 top liberal arts colleges, with their dominant focus on undergraduate education geared to citizenship, be especially engaged in this process?

Although some research in these general areas has been conducted, rarely has the resulting information been applied to the policy realm of consumption and leisure time. Opportunities exist, then, for linking researchers working on student attitudes[28] to those reflecting on how to frame a low-consumption agenda. Likewise, several practitioner communities exist with deep knowledge of linking student groups to transnational advocacy networks. But here too, their knowledge isn't easily or applied to the TBYT initiative. These practitioners aren't documenting their "lessons learned" in accessible ways, and no researchers have stepped forward with a systematic analysis that would fill the gap.

---

[27] See, for example, the India Resource Center's Coca-Cola resources at www.indiaresource.org/campaigns/coke/index.html

[28] For example the Higher Education Research Institute at the University of California Los Angeles at http://www.gseis.ucla.edu/heri/heri.html

Research Arena Three: Cementing an Environmental Politics of Time Famine

For at least two important reasons, major environmental NGOs in the US have been slow to incorporate "time politics" into their educational and policy agendas. Doing so would have diluted their core message of environmental protection during a time of unusual government hostility toward environmental protection. It may have also alienated supporters for whom the connection between time famine and overconsumption is difficult to see. However, as issues of environmental well-being become increasingly linked to the dynamics of consumption, US environmentalism must become more open to confronting the fundamental drivers of overconsumption. TBYT's connection of "vacation rights" to eco-travel and nature appreciation is a first, critical step toward cultivating such openness.

Moving beyond this first step won't happen easily or automatically. Recent voices within the US environmental community argue, for example, that "Apollo project" programs to develop new technologies of production and consumption must occupy the center of any move towards contraction and convergence. As tantalizing as these Promethean possibilities might be, they divert attention from the drivers of consumption, and of the ways in which structural change in work–leisure arrangements can slow the maddening treadmill of work and spend. If mainstream environmentalism is to stay focused on the connections between overwork and overconsumption, it will need considerable help from the research community, in at least the following three ways:

- *Building on "Vacation Rights"* The "Right2Vacation" initiative argues that more paid vacation time will lead to lower work stress, reduced binge vacationing, higher levels of local civic participation, deeper connection to (and appreciation of) local and regional environmental assets, and a growing political awareness of the benefits of trading income (and consumption) for leisure. These arguments are plausible on their face and enjoy some empirical support. Supporting research, however, is spread across several disciplines, dated, ill-matched to contemporary environmental concerns, or insufficiently robust to inform or motivate ambitious policy commitment by major environmental groups (and other political actors). There are significant opportunities, then, for the research community to synthesize and extend existing knowledge about the impact of extended paid vacation on consumption, travel, and the cultivation of civic and environmental sensibilities. This work could begin with a review of the varied literatures to develop a "state of knowledge" overview and assessment. Further work might explore the interplay between additional vacation time and environmentally optimal outcomes, or identify mechanisms for framing or institutionalizing vacation time in ways that foster high-leisure, low consumption activities.
- *Conceptual Brush-Clearing Regarding "Sacrifice"* Do some kinds of reductions in material consumption yield increased happiness, while others do not?

Probably so, but talking easily and naturally about these two categories proves difficult in a political and linguistic environment that reflexively equates all consumption reductions with dire sacrifice. Lacking are clear conceptual frameworks and an everyday language, supported by compelling everyday examples, that would allow policymakers and environmental groups to easily distinguish (for themselves and a sometimes skeptical public) reductions in material throughput that are happiness expanding from those that are not. Right2Vacation and TBYT are experiments in developing this sort of language – but these efforts remain less than intuitive, and their power over the popular vernacular of environmentalism remains unclear. What sorts of language and frames best convey the possibilities of reduced consumption in service of human happiness?

- *Animating the "Base"* TBYT and Right2Vacation are policy extensions of the voluntary simplicity movement. In some ways, both initiatives should have taken off long ago. After all, the available data suggest that at least a quarter of Americans are fundamentally sympathetic to notions of voluntary simplicity and time famine. The dilemma is that this base group of simplifiers sees political change as a function of individual acts of frugal consumption rather than the coordinated exercise of citizen power (another example of this can be found in Chapter 3, this volume). How can this group be "turned" toward a deeper engagement with citizen action, in support of TBYT's agenda? That's a surprisingly difficult question to answer. There has been scant systematic assessment in the last decade of public attitudes toward simplicity and entry points for fashioning action coalitions within this population. Little is known about the groupings and composition of key social and culture groups, in either (or both) the global north and south, that may be most receptive to a message of consumer restraint, and thus most readily enlisted in a political program of policy change. The largest marketing-research organizations probably have some of this information; one research task, then, for any drive – national or transitional – toward a global norm of consumer restraint is to discern how to leverage these data. Another task is to develop a rough data base of the many research endeavors[29] aimed at identifying those global constituencies most undermined or diminished by time famine and the decline of leisure time and civic consciousness. Perhaps by bringing together, in crude analytic ways, the conclusions and data of these myriad groups, important patterns will emerge that will facilitate a networking of key groups around the world *and* a joint identification of critical, and perhaps counterintuitive, constituencies.

[29] For example Harvard University's Project on Global Working Families at www.hsph.harvard.edu/globalworkingfamilies/

The Challenge to the Sustainable-Consumption Research Community

The story of TBYT and its partial metamorphosis into a "vacation rights" initiative offers three critical lessons for the research community. The first is that the natural-science culture of polite discourse and the ensuing retreat from the political[30] that characterize much of the scholarship on sustainable consumption may poorly serve activists who would advance a research-informed agenda for sustainable consumption. Like many of us working in this field, TBYT's first impulse was to be inclusive, accommodating, and open to multiple interpretations and actions. Appearing "reasonable," after all, is thought to confer credibility, and with credibility comes effectiveness. But the story didn't play out this way. Only when TBYT sharpened its message and spoke in sometimes uncompromising ways did genuinely viable opportunities for collaboration, coalition building, and public-policy pressure emerge. Scholars of sustainable consumption who would see their research products translated into effective policy outcomes may do well to reflect on the lessons of this case. While responsible science and collegial respect are always to be valued, there are costs to being reasonable, polite, and inclusive to a fault in the face of mounting environmental threats to human well being.

The second lesson is simply that there are a rich set of policy possibilities for fashioning political and economic arrangements that can enrich human satisfaction while reducing consumption. The TBYT experiment is just one of many. Rather than lament the lack of "political will" to confront the costs of unsustainable systems of production and consumption, researchers might begin investigating the ways in which political will might best be forged. For this, TBYT offers a third lesson: unlikely coalitions and hidden stakeholders in a politics of sustainable consumption lurk just beneath the surface. Identifying and mobilizing these actors may depend less on the comprehensiveness of statistics and indicators, and much more on the framing of issues. de Graaf is employing essentially the same set of objective facts in his "vacation rights" program as he was in his earlier TBYT efforts. Indeed, he uses the vacation initiative to speak more broadly to diverse audiences about the need for a cultural shift that would trade productivity for free time rather than consumption. The framing, however, is different, and out of this difference has emerged a coalition of disparate actors – students, travel agents, progressive corporations, environmental groups – that may yet prove successful in injecting time politics into national debate.

The math is inescapable. The planet cannot sustain nearly 7 billion people living the American (or even European) way. Happily, it doesn't have to for most of the world's inhabitants to enjoy a prosperous and satisfying life. Shifting from a global norm of "more is better" to one receptive to low-consumptive paths to happiness

---

[30] On the history and political costs of this emerging ethos of "politeness" in mainstream environmentalism, see Mark Dowie, *Losing Ground: American Environmentalism at the Close of the Twentieth Century*, Cambridge, MA: MIT Press, 1996.

and security must rise to the top of any agenda for sustainability. The challenge, however, is to engage critical actors and significant slices of the public in a politics of structural change that would make, in the words of Paul Hawken "acting sustainably as easy as falling off a log."[31] Long relegated to the margins of environmentalism, and trapped by its own predisposition to individual consumer actions that consciously avoids broader collective action, the "simplicity movement" in the United States now struggles for political salience and public-policy power. If the movement finds its political footing in the US – arguably the epicenter of contemporary consumer culture – it becomes difficult to ignore the political feasibility of tying consumer restraint to increased prosperity. The door to a vibrant transnational politics of "how much is enough" then swings open a bit more. By delinking human happiness from material provisioning, recent research points to an affirming politics of "de-consumption" that could bolster human well-being without invoking the politically paralyzing specter of sacrifice. The next step is to build on such research with inquiry and insight around the networks, organizing dynamics, critical actors, and social groupings that together could propel the time–famine movement onto the US, and thus transnational, stage[32].

---

[31] From the preface of Paul Hawken, *The Ecology of Commerce*, New York: HarperCollins, 1994.

[32] Portions of this chapter are drawn, in revised form, from Michael Maniates, "Struggling with Sacrifice: Take Back Your Time and Right2Vacation.org," in Michael Maniates and John Meyers eds., *The Environmental Politics of Sacrifice*, Cambridge, MA: MIT Press, 2010.

Chapter 3
Sustainability Transitions Through the Lens of Lifestyle Dynamics

Fritz Reusswig

Reducing the Carbon Footprint of Modern Consumer Societies

Satisfying an ever-increasing consumer demand puts a strain on the environment as increasing amounts of space, material and energy are needed (Guinee 2002; Princen et al. 2002; Redclift 1996; Stern et al. 1997; Shove and Warde 1998). Material intensive consumption is accompanied by increasing amounts of waste and emissions. According to our own assessment, about 19 million tons of industrial CO_2 emissions (25% of the total) can be attributed to direct lifestyle and consumption related human activities, most of which occur in the industrialized world, but with a growing share from rapidly developing countries such as China or India (Reusswig et al. 2004).[1] Despite their still (very) low level of material consumption and related emissions on a per capita basis, the total carbon footprint of these emerging economies has reached significant levels.[2] Due to catch-up processes and globalization effects, the dynamics and the environmental effects of modern consumer society is no longer confined to its historical region of origin: the United States (US), Western Europe, and – more recently – Japan. Economic growth, political modernization and cultural globalization do not only lead to the *overall* growth of resource use and emissions, they also change the *internal* composition of societies. Myers and Kent (2003) account for 1,059 million additional people having joined the global consumer class. This includes the expanding middle class in countries such as China, India or Brazil (Bhalla et al. 2003; Consumers International 1997; MGI 2006, 2007;

F. Reusswig
Potsdam Institute for Climate Impact Research, Potsdam, 14412, Germany
e-mail: fritz@pik-potsdam.de

[1] If indirect (induced) resource flows and emissions are included, the environmental impact of consumption is even higher (Hertwich et al. 2005), contributing substantially to the total 'ecological footprint' of a society (York et al. 2004), or its metabolism (Fischer-Kowalski and Amann 2001).

[2] China has already passed the US as the largest emitter of greenhouse gases. The rapid growth of overall emissions will also affect the accumulated emissions over time. While the US, Europe and Japan are still 'leading' in overall emissions, this will change in the near future when China is expected to overtake the U.S. in 2021 and India will keep up with Japan 10 years later (Botzen et al. 2008).

L. Lebel et al. (eds.), *Sustainable Production Consumption Systems: Knowledge, Engagement and Practice,* DOI 10.1007/978-90-481-3090-0_3,

Robison and Goodman 1996; Sridharan 2004; van Wessel 2004). Global studies show that the propensity to consumerism and the associated dreams and hopes – often fuelled by advertising and other global mass media products – of the emerging consumer class fuel future production and consumption processes especially in countries with a higher proportion of poor people (Environics 2002).

At the same time, the Fourth Assessment Report of the Intergovernmental Panel on Climate Change (IPCC 2007) has made it clear that the current trends in emissions and subsequent anthropogenic climate change require immediate action both to enhance the adaptive capacity of societies as well as to mitigate against the causes of global warming. A widely held consensus among climate scientists – and a policy goal adopted by the European Union – states that the global community should try to prevent 'dangerous climate change' by limiting a Global Mean Temperature (GMT) increase of +2°C against pre-industrial levels (Schellnhuber et al. 2006; Walker and King 2008).[3] For a high-level emission country such as Germany such a goal would translate to emission reductions of about 80% (base year 2007) in order to meet a global environmental and equity goal in 2050.[4]

Such drastic reduction goals cannot be achieved by some energy saving light bulbs here and a few miles of energy efficient car there. These require a Sustainability Transition (ST) in modern societies and of production and consumption systems in particular (Lebel 2005). By 'Sustainability Transition' (ST) (cf. NRC 1999) I refer (1) to a *normatively* influenced, yet fact based concept of how humankind will have to evolve in order to meet their needs and wants based on intra- and inter-generational equity criteria and without dangerous interference with the Earth's ecosystems. The broad notion of 'Sustainability Science' (Kates et al. 2001), encompassing different research domains and social discourses (e.g. in the UN system), might be regarded as a focal point for the type of science required for the concept and its further evolution.

ST refers (2) to an *ongoing* social process that tries to assess and realize the viability of this concept by building and creating sustainable systems, technologies and social practices that are more or less intentionally inspired by the idea of a change of existing more or less non-sustainable practices and structures. Here the concept has clear links to ecological modernization theory (Buttel 2003; Huber 2000; Spaargaren 2003).

The literature on ST (Kemp et al. 1998, 2007; Kemp and Loorbach 2003; Rotmans et al. 2001) offers at least four advantages for a social science oriented view: (1) it takes a long term perspective, and addresses issues of timing, (2) it has a focus on complexity and multi-level character of change, (3) it has a clear focus on technological change, and (4) it explicitly addresses issues of management and governance, which offers a fresh alternative to the widespread sociological attitude of pure observation.

---

[3] The measured increase in GMT is about +0.76°C; additional +0.6°C are already 'in the pipeline', but have not materialized yet due to inertia of the Earth system. The two-degree goal thus leaves us with a very limited window of opportunity for reducing emissions.

[4] Per capita emissions of greenhouse gases in European countries are about 10 t per year, while they amount to almost 20 t in the US.

As the ST literature has not yet explicitly addressed lifestyles and lifestyle changes, this chapter wants to highlight the necessity to link technology and governance issues by taking lifestyle changes into account. We will not be able to meet global sustainability goals without asking questions like 'Do we need this?' or 'How much is enough?' (Durning 1992). Following this intuition, the rest of the chapter is organized as follows: I would first like to embed consumption into a conceptual framework defined by lifestyles and how they influence wider social changes. In a next step, I would like to illustrate one dimension of lifestyle dynamics empirically using two cases: wind energy development in Germany and the case of Product Carbon Footprint. In the next section, a more general perspective will be taken, looking at the change of the lifestyle composition of a society and the diffusion of pro-environmental attitudes and behaviors. The conclusions are the final section.

From Consumption to Lifestyle

'Consumption' means different things to different people. Scientists interested in material flow analysis, often influenced by physics, engineering or biology, refer to the use or flow of materials and energy through a bio-physically defined system. The advantage of being physically explicit and environmentally revealing is here usually traded against the lack of understanding of economic processes and social embeddedness. When economists talk of consumption, they usually refer to (private household) purchases of goods and services on markets. Preferences, prices, quantities, and market equilibriums are dominating focal points of the economist's attention, but physical flows and environmental impacts usually get lost. Ecological economics tries to overcome this deficit (Duchin 1998; Røpke 1999, 2005). Ecological economists challenge basic assumptions of the neoclassic tradition (such as the neglect of external costs), and sometimes even address the post-purchase phase (Cogoy 1999). Sociologists, who have been neglecting consumption due to the production bias of their discipline for quite a while, usually refer to the symbolic processes by which social actors express and reproduce social inequalities and cultural values. While some scholars focus more on the subjects of consumption, others address the meaning of consumption in (post-) modern societies as a whole (Schaefer and Crane 2005; Zukin and Maguire 2004).

In this chapter, 'consumption' refers to the processes of preference formation (e.g. via advertising and public communication), purchase, use, and disposal of goods and services by individuals in private or corporate households (organizations, governments) in a social context. Consumption is linked in complex ways to production processes, distribution structures, and other systems of provision, and combines material/energy and social (symbolic) aspects.

Consumption processes are embedded in lifestyles. The term 'lifestyle' is widely used in environmental (and even in sustainability) contexts. Many scholars, and even some practitioners, underline the need for changing 'our lifestyle'. But given the plurality of lifestyles, and lifestyle concepts, in modern societies, the question arises: who should change in what direction? In order to avoid confusion, at least three levels of analysis can be distinguished here (Table 3.1).

Environmental sociologists – and, of course, psychologists – have often looked at (pro-) environmental attitudes and behavior changes (and barriers to them) at the micro-level of individuals and households. At this micro-level, the term 'lifestyle' refers to individual ways of leading one's everyday life; this sociological tradition can be traced back to Max Weber and his analysis of the 'Protestant Ethic'. The bulk of literature on (sustainable, pro-environmental) attitudes and behavior can be related to this level.

At the macro-level of society as a whole, 'lifestyle' refers to typical behaviors and mentalities, influenced by the network of social interaction and average living conditions (e.g. technological infrastructures). The often quoted 'American Way of Life' would be an example for a macro-level lifestyle type. Environmental sociologists have been looking at this overall level of social performance (e.g. Uusitalo 1986) and change (e.g. Reusswig et al. 2004).

At the meso-level, the term 'lifestyle' refers to patterns of activities (usually in consumption and leisure), to associated attitudes and values, and to characteristics of the social situation of groups of individuals. There is no commonly shared definition of the concept in sociology. Major debates about the methodologies to be used, and the role of the lifestyle concept in comparison to class occur: following Pierre Bourdieu (1976) many sociologists tend to see the consumptive and expressive side of a lifestyle more as an expression of social class and its related forms, whereas sociologists like Beck et al. (1999)) and Urry see lifestyles more as autonomous expressions of individual choices, independent of class.[5]

My consequence would be to define lifestyles in an inclusive way, taking three main dimensions into account: *social structure*, defining the resource endowment

Table 3.1 Levels and dimensions of lifestyle

Level of analysis	Definition of lifestyle	Ontological reference	Main methodologies	Lifestyle dynamics
Micro	Individual ways and forms of everyday life	Individual	Narratives, observation, qualitative interviews	Biographical and intra-generational changes in attitudes, practices, and habits
Meso	Group specific patterns of leading and interpreting individual lives	Group	Qualitative and quantitative surveys, factor and cluster analysis	Changes in social capital and social structure
Macro	Typical behaviors and mentalities of societies	Society	Cultural studies, mentality history, macro-sociology	Transition of mentalities, technologies, and infrastructures

[5] For a brief summary of this debate see Tomlinson (2003).

and constraints, *performance*, circumscribing the practical and expressive side of lifestyles, and *preferences*, catching the evaluative and motivational side (cf. Lüdtke 1989; Müller 1992) (Fig. 3.1).

Lifestyles, defined at the meso-level, are group specific forms of how individuals live their lives (performance) and interpret them (preferences). They imply questions of (social) identity and meaning. Lifestyles result from individual choices under social constraints (structure). Consumption is both an expression of and a resource for lifestyle formation/reproduction.

In modern societies, there is not one typical lifestyle, but a plurality of different lifestyles, and this differentiation is important for a ST both at the descriptive and the normative level (Reusswig 1994). Lifestyles at this social meso-level have attracted less attention from environmental sociologists. However, an impressive body of literature has emerged from the areas of social structure analysis and cultural sociology. In addition, commercial market and media research provides us with valuable information about market segmentation in modern societies, often specified for different kinds of products and services. Studies that have looked into internal differences of modern lifestyles with regard to resource consumption and emissions reveal significant differences. Lutzenhiser and Hackett (1993) for example found factor four differences between high and low household CO_2 emissions in urban U.S. households. A similar study for European households detected factor three differences (Weber and Perrels 2000). If 'green lifestyles' are explicitly included in the sample, differences are even larger: Christensen (1997) found factor 8 differences between the lowest and the highest emission families ('American Lifestyle') in Denmark.

When the analytical focus is moved from 'consumption' to 'lifestyle', consumption becomes transparent as a socially and culturally embedded process. Consumption

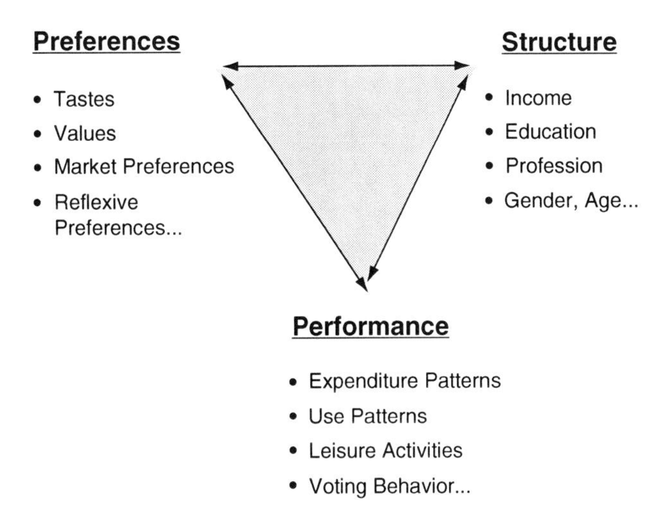

Fig. 3.1 Main dimensions of lifestyles at the meso-level

processes and their environmental consequences can be studied in the context of social inequalities, cultural traditions, and infrastructural boundary conditions.

Scholars of modern consumer society disagree about the role that consumption processes (such as preference formation, shopping, using consumer goods, communicating about theme, etc.) are playing in modern societies. While some argue that consumption is more or less an appendix to the production system in modern, capitalist societies (e.g. Fine and Leopold 1993), others highlight the structuring power of consumer's preferences and consumer culture over production (e.g. Campbell 1987).[6]

There is evidence for both lines of thought, and I do not see any reason why environmental sociology should not accept complexity and multi-causal relations. The important point is the specification of the routes of influence in their nature and, if possible, their strength. If the conceptual focus is moved from consumption to lifestyles – and if the material and environmental aspect is not lost – this multi-causal and socially connected character of modern consumption processes gets higher visibility. In addition, the dynamics of consumption processes is better conceivable. In order to elaborate the latter point, I would like to illustrate the complex causal relations between social institutions, technology and lifestyles in a simplified form (Fig. 3.2).

I distinguish three basic elements: (A) social institutions and values characterize the major driving forces and determinants at the social macro level, (B) products, technologies and systems of provision refer to the physical 'fabric' of a society, including its organizational structure, and (C) lifestyles refer to individual ways of

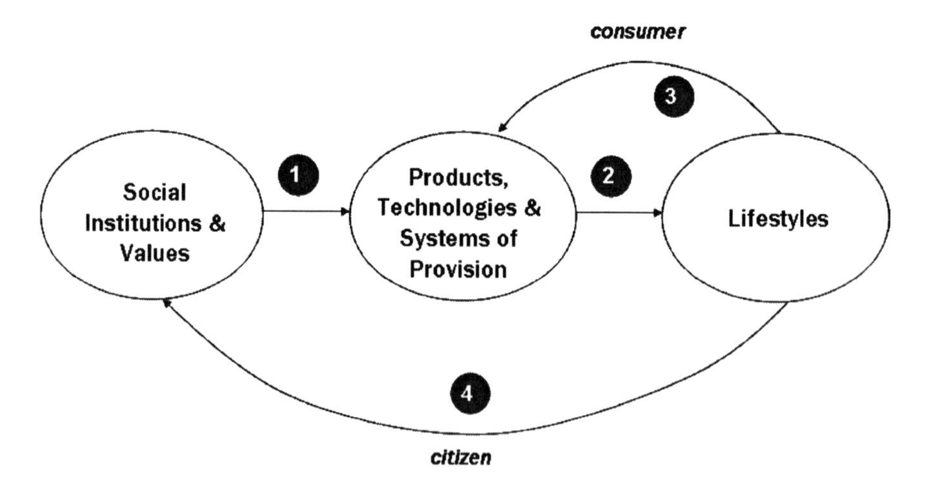

Fig. 3.2 Simple model of the role of lifestyles in the context of social institutions and values and products, technologies and systems of provision

[6] For an overview, cf. Zukin and Maguire (2004). This debate has clear consequences for environmental sociology addressing consumption issues: if the production advocates are right, every attempt to achieve more sustainable consumption via the consumption side is futile.

leading and interpreting ones life – be they defined purely individualistic, or as group specific patterns. The straightforward arrows indicate two things: (1) products, technologies and systems of provision are influenced by social institutions and values, and (2) they in turn influence (enable, constrain) the behavior (practices, routines) and, to some degree, the attitudes of (individual) consumers. Relation (1) can be illustrated by studies that demonstrate how product, technology and systems development are by no means autonomous processes steered by engineers, business organizations and municipality organizations, but heavily influenced by underlying social institutions and their guiding problems, interests, and values. Cases in point here are research streams such as National Systems of Innovation (Lundvall 1995; Porter 1990), Varieties of Capitalism (Hall and Soskice 2001), and History of Technology studies (Hughes 1983; Stier 1999).

Relation (2) has been repeatedly highlighted by environmental sociologists (sometimes even psychologists) hinting at the limits of individual (consumer) behavioral changes as drivers of a Sustainability Transition. Diekmann and Preisendörfer (2003) stress that environmental sociologists should take the costs of individual behaviors into account, and they highlight how social and material infrastructures influence cost structures. Spaargaren (2003) has pointed out how closely individual lifestyle and consumption changes are linked to (and constrained by) systems of provision. His work is influenced by Cowan's (1983) analysis of the effect of infrastructural systems on the labor of private households, especially women. Shove and Warde (1998) have, in much the same spirit, highlighted the crucial interplay between systems of provision and the routines and standard expectations of everyday life.

With regard to short term changes in consumption behavior it might be tolerable to neglect these structural effects indicated by the relations (1) and (2) in my simple conceptual model. With regard to long-term and system-wide changes, however, environmental social scientists more or less involuntarily promote naïve, if not ideological research agendas by doing so. Given the complexity of social conditions for individual consumption, there are clear limits for burdening the individual consumer with the whole array of system wide changes. Other actors, such as the business sector, civil society, and governments clearly have to play their roles as well (Jackson 2004, 2005). A closer look at the field of consumer policy reveals clear deficits and disincentives, often being much more influential on individual consumer behavior than the widely accepted rhetoric of 'sustainable consumption' (Fuchs and Lorek 2005).

It would be naïve, if not ideological in almost the same manner, however, if environmental scientists and activists would confine themselves to these structural effects on individual action exclusively. The rich body of literature on social change reveals a dialectical relation between individual action and social structure (Hernes 1976). For social institutions and for technologies Giddens's 'Theory of Structuration' (Giddens 1986) holds as well: structures enable and constrain individual action, but individual action influences social structures. Two basic ways of individual action influencing structure should be distinguished when thinking about consumption and lifestyles: market preferences, expressed by individuals and households as *consumers* (3), and reflexive and policy preferences expressed by individuals as *citizens* (4). This double role has been illustrated by research findings about ethical consumption, consumer policy, and corporate dialogues with consumer-citizens (Carrigan et al. 2004; Cohen et al. 2005; Crocker and Linden 1998; Doubleday 2004; Manoochehri 2002; Stevenson 2002).

Ulrich Beck has termed the consumer 'a sleeping giant'. His or her giant-like abilities become visible only if we take both the consumer and the citizen into account. For reasons of brevity I would like to confine the influence of the consumer to products, technologies and systems of provision, whereas the citizen mainly influences the boundary conditions, that is institutions and values. The next sections elaborate on the influence of citizens and consumers on technologies and social institutions. Their purpose is not to neglect the influence of the latter on lifestyles, but to highlight how lifestyles and lifestyle changes contribute to overall social change – including a ST. Two examples will be discussed in more detail: the rise of wind energy in Germany and the introduction of product carbon footprints. In both cases the interplay between consumption and lifestyle changes, technological change and institutional dynamics (including politics) is essential. In the following sections, I will try to embed these changes into the wider context of social change in general.

The Success of the German Wind Energy Sector

The necessities to reduce the carbon footprint of the German economy have early on been translated into debates about its future energy mix, including timing and costs. Most experts agree that renewable energy carriers (wind, solar, biomass, etc.) will have to play a central role here. Questions are more 'how', 'when', and 'at what costs'. At the turn of the century, it has become evident that we might well witness the initial phase of the required energy transition, with renewable energy growing rapidly – but from a rather low absolute level. The German wind energy sector displays remarkable growth rates. It has outperformed early leaders like Denmark and the U.S. since the 1990s. One of the major reasons for Germany being so successful in implementing wind energy is a political one. In 1991 and 2000, two federal legislation acts (the Feed-In Law and the Renewable Energy Law) have spurred growth substantially. International comparisons show how the German model of guaranteed tariffs, for a limited period subsidized by electricity consumers, ensures investment security for wind energy providers and has proved its superiority to the quota model favored by countries like the UK or Italy (Bechberger and Reiche 2005, 2006).[7]

---

[7] It is worth noting that wind energy alone will of course not suffice to provide sufficient electricity for the economy of a highly industrialized country. Other renewables will have to step in too. Photovoltaic and solar thermal systems have a smaller market share, but show impressive growth rates in Germany, covered by the Renewable Energy Law. The future of the transportation sector is open, as many options are followed (electricity, fuel cell, hybrids, agrofuels, and methanol). For a transition period, even the option of Carbon Capturing and Sequestration (CCS) is a climate neutral possibility, buying time for the renewables to gain momentum. The future of wind energy will very much depend upon the development of offshore facilities, which has just begun.

However, the legislative aspect is only part of the success story, and it has to be seen in a wider context – a context of a social–ecological transformation of the German society, in which lifestyle dynamics plays a constitutive role. It was the interplay between the political system and the energy economy on the one hand, and lifestyle dynamics on the other that created what might be termed the take-off phase of the German wind energy sector. In the lifestyle dynamics domain, both the 'consumer' and the 'citizen' have been important.

Looking back to the 1970s, the German economy was mainly based on fossil fuels plus a rising share of nuclear power. The energy crises of the 1970s led to some governmental research in renewable energy systems, but the main effort was to diversify the geographical distribution of oil resources, and to construct more nuclear power plants. After 1968, the Federal Republic of Germany – like many other countries during that period – experienced a political shift to the left, the growth of left-wing, more or less radical political parties and groups, and the rise of an 'alternative milieu', that is a small group of – mostly young and well-educated – people, striving for a more 'natural' way of life, including private consumption patterns and 'green' forms of production and living together. These 'eco-pioneers' were strictly opposing big industry and government, often living in social (and geographical) niches of the mainstream society. As the federal government propagated nuclear power as a solution to the energy crises of the 1970s, these groups began fighting against government policies, mostly peacefully but time and again quite violently.

The critical, catalyzing event for this whole constellation was the Chernobyl accident in 1986, which substantially de-legitimized nuclear power in Germany.[8] Many proponents of the German environmental social movement felt the urgent need to do something concrete and constructive, and the citizen's wind energy movement was born (Byzio et al. 2002). Groups of engaged citizens organized under the umbrella of cooperative societies under German public law, allowing for risk sharing, and imported Danish wind turbines to be erected in their backyards. If we take Max Weber's typology of social action into account, their activities were driven by value rationality (political goals, idealism) and emotional rationality (fear of nuclear disaster). Especially in Northern Germany, where wind conditions are favorable, this citizen-based wind energy movement gained momentum during the late 1980s, including the emergence of expertise, small businesses, and pressure groups.

[8] By coincidence, 1986 was also the year when the public debate on climate change gained momentum in Germany (Weingart et al. 2000). As nuclear power generation is associated with much less CO_2 emissions than coal, oil or gas powered plants, the rise of climate change as a major issue in the environmental discourse could easily have led to a strengthening of the pro-nuclear option – a route that many nuclear power advocates clearly intended to take. Nevertheless it did not work out: the majority of the German public (and, in particular, the 'alternative milieu' of these days) was concerned about climate change (in the media often termed 'climate disaster', *Klimakatastrophe*), but at the same time remained deeply skeptical about the risks of nuclear power. For German environmental sociologist Ulrich Beck the year 1986 was crucial as well: the first edition of his 'Risk Society' was published immediately after the Chernobyl event, offering unprecedented public resonance to a work by a sociologist.

It is important to observe how the federal government followed a different route. Under public pressure to limit or abandon nuclear power, research and development expenditures for renewable energy sources grew slightly during the 1980s. Together with some large energy providers, the GROWIAN project, a research and test facility, was built in northern Germany in 1983.[9] The administrative and scientific preparation of a large windmill test facility had taken place from 1973 to 1979, initiated by the first 'oil crisis'. Scientific wind turbine experts were able to influence the process by assuming basic problems to be already solved, and by advising the government to go for a big technological solution. GROWIAN's dimensions (hub height: 100 m, rotor blade diameter: 100 m, 3,000 MW) have been reached by commercial wind turbines only at the beginning of the twenty-first century. The coalition of government officials, wind energy experts, and big business representatives clearly wanted to realize a 'big solution', dwarfing contemporary Danish and US developments. However, this 'big leap forward' failed, GROWIAN was running only a few hours per day in total.

At the same time, the citizen's wind energy movement followed technologically a much more 'conservative' pathway, using well-established and low-risk technologies, and engaging in a gradual learning-by-doing process. Despite their far-reaching energy policy goals, their everyday practice was governed by a moderate, stepwise approach, facilitated by economic and social constraints, such as availability of financial means, pooling of resources with people with a similar lifestyle, building-up of expertise, problems with local authorities, etc. This resulted in a small-scale niche of a politically inspired production–consumption system.

1991 was a crucial year for wind energy in Germany. The European Union had started to think actively about the liberalization of energy markets, and oil prices were very low. The wind energy lobby, which had emerged by that time, used the window of opportunity and teamed up with the German hydropower lobby in pushing the federal government for a feed-in tariff system. Despite some opposition from the big energy providers (which were engaged in 'swallowing' the East German energy system after reunification in 1990), the law was passed by a conservative/liberal government. It provided economic security for wind energy providers, reducing their investment risks, and attracted new groups of consumers and investors. Due to the changed incentive structure, other social groups with different motivational and attitudinal backgrounds became interested in wind energy. The 'necessity' for politically motivated idealism, so indispensable during the pioneering phase, vanished. It gave way to a much more pragmatic attitude and even profit-seeking behavior. In Weber's terms, this was a shift from value rationality and emotional rationality to purpose oriented rationality – and after a time even other new forms of traditional rationality.

The organizational structure in the production–consumption system of energy provision changed accordingly. At the beginning, a rather informal design was chosen (cooperative society), suited for networks of people with relatively strong

[9]GROWIAN is the abbreviation for *Grosse Windenergie-Anlage* (Large Wind Energy Facility).

ties, a low degree of specialization, open for idealism and private engagement, appropriate for risk pooling and a minimal degree of formal communication with the outside world (mainly public authorities). The governance principle can be circumscribed as network or solidarity. Today, after the two major legislative achievements, the market has taken over. Many of the citizen's wind power organizations still do exist. However, the majority of the capacity growth since the 1990s has been achieved by medium and large sized joint-stock companies. Shareholders are anonymous, and the degree of specialization is high. Internally, more formal hierarchy is in place.

The legal boundary conditions for wind energy generation in Germany after 1991 were well in favor of their growth, fuelled by a 'normalization of ecology' (Brand et al. 1997), instead of the idealism of the early years.

> Legitimacy and visions are shaped in a process of cumulative causation where institutional change, market formation, entry of firms (and other organizations) and the formation and strengthening of advocacy coalitions are the constituent parts. At the heart of that process lies the battle over the regulatory framework (Jacobsson and Lauber 2006: 272).

However, the citizen's wind energy movement of the 1980s, a clear descendant of the environmental movement and situated within the alternative milieu, was crucial for the success story of German wind energy in at least four ways: (1) by creating a domestic market for small wind energy systems, the movement helped to provide demand, experience, and cost reduction. (2) By helping to constitute a small (domestic) industry, the movement contributed to the formation of an advocacy coalition (including political party members and scientists) that was able to actively influence the legislative process. (3) By helping to create human capital around the construction and maintenance of wind turbines, the movement contributed to the nurturing of professional nuclei for further development in universities, the business sector, and the administration. (4) By combining far-reaching energy policy goals with a good deal of pragmatism and the ability to form coalitions with otherwise opposing groups, the movement paved the way for a more encompassing social consensus on renewable energy as a necessary and viable option for Germany.

If we look back at the process from a bird's eye point of view (cf. Fig. 3.3), we observe not only phase shifts – from nuclear consensus and (soft) exclusion of alternatives in the 1970s to the promotion of renewable energy and the nuclear exit option in the 2000s – we also observe changes at the lifestyle and civil society level. The important point, however, is that these two domains do not operate in isolation, but are connected with, and influence, each other. The arrows no. 1 and 2 in Fig. 3.2 tell one important part of the story: overall social values and political decisions shape technologies and infrastructures, and technologies and infrastructures influence our consumption patterns and lifestyles. However, the arrows no. 3 and 4 hold as well: consumers shift demand patterns, creating the need for technology changes, and citizens influence the value structure and the political process of a society, which in turn affects the technological structure. The point I wanted to make in this chapter is not to say that the citizen's wind energy movement was 'causing' Germany's international success in wind energy performance. I simply wanted to illustrate a

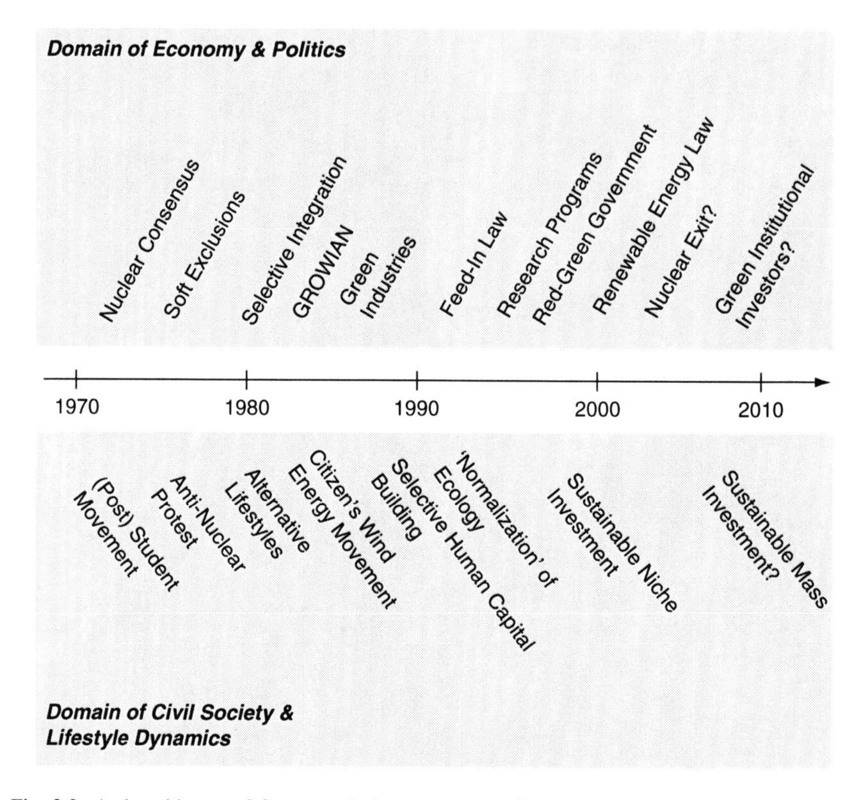

Fig. 3.3 A short history of German wind energy system in the interacting domains of economy and politics (*top*) and civil society and lifestyles (*bottom*)

multi-causal network of social and technological change, with consumers and citizens being active parts, not only passive recipients of a sustainability transition.

In order to become really sustainable, the world's energy system has to change dramatically. The issue of energy poverty has not been touched by this chapter, but should be mentioned. Given the relatively small size of European societies, the question of future energy sustainability will be answered mainly elsewhere, in China or India for example. In these countries growth rates for renewable energy are impressive (as are other growth rates). And in part these rates are triggered by a network of interactions, including technology exports and international environmental agreements. However, we must not forget that consumers and citizens do play a role in these countries as well (Consumers International 1997). Membership in civil society organizations that is oriented towards environmental issues is growing, and it grows primarily in middle income countries (Anheier et al. 2001).

There is no reason to be overtly optimistic. But there is reason to go beyond a too narrowly defined image of consumption and the consumer, leaving the citizen aside. The socio-economic processes that are increasing the number of consumers in China and India also contribute, although not automatically, to higher expectations of environmental quality and political participation.

Product Carbon Footprints: A New Tool for Systemic Innovation in Producer–Consumer Networks

About 40% of total GHG emissions in industrialized societies arises from consumption processes and consumer goods, including food and beverages. Given the complexity of today's production–consumption networks, often organized across the globe, the reduction of the carbon footprint of consumption cannot be met by the group of consumers (private households) exclusively. System-wide solutions and new institutional settings (including incentives and regulations) are necessary. One key aspect, at least from a consumer point of view, is the lack of transparency. Both national and global polls on climate change perception show a widespread concern about global warming, combined with a (somewhat reduced) awareness that individual action is necessary. However, due to a lack of (credible) information on the carbon footprint of consumer goods and consumption processes, people cannot translate their concerns as citizens into everyday choices as consumers. GHG emission assessments across the total lifecycle of products exist, but they are rare, methodologically not standardized, and lack consistent public communication. Single aspects or terms such as 'food miles' sometimes emerge and dominate debates, while other aspects such as agricultural inputs to food products or energy savings due to insulation material are widely neglected – or circulated only occasionally. As a result, consumers – and producers, although to a lesser degree – remain confused.

Given the necessities of climate policy, the world community cannot afford to remain confused and thus disabled as consumers. One important step to overcome confusion and inaction has been taken by UK's Carbon Trust, which introduced a carbon label for consumer goods (Carbon Trust 2007). Potato chips and hair shampoo have been among the first products labeled, but UK's largest retailer Tesco has announced that it would label all its products on the shelf in the coming years. The debate on carbon labels has started all over Europe (cf. Migros in Switzerland or Leclerq in France). In Germany, a consortium of research and policy organizations (including Oeko Institute, Thema1, WWF Germany, and Potsdam Institute for Climate Impact Research) has initiated a pilot project together with ten large companies (e.g. Tchibo, dm, Deutsche Telekom, Rewe *Group*) (cf. http://pcf-projekt.de/main/news/?lang=en). The main goal of that pilot project is to develop a common and scientifically sound methodology for product carbon footprints (PCF), and to strive for international standardization.

On average, every single German consumer is responsible for about 11 t of CO_2 eq. per year. Four tons (or 36%) are related to food, drinks, and other consumer goods. This is more than the building or travel-related emissions respectively. PCFs would be a key instrument to raise consumer awareness, and to create carbon transparency in the domain of consumer goods which actually is either lacking – or distorted by single issue foci such as food miles.

Given the global character of many value chains in today's economy, the shift towards constant carbon dioxide emissions reductions at one point affects all others, including extraction, manufacturing, transportation, and waste disposal.

It is clear that PCFs alone will *not* pave the way to a low-carbon economy. On the other hand, PCFs can become one core element of such a structural shift. From an innovation-oriented point of view, there are some very interesting aspects of a PCF.

- PCFs are situated at the interface of production and consumption. They reflect the total value chain related emissions, and communicate them to the final consumer. A CO_2 label will enable consumers to translate their concerns as citizens into consumption choices, while at the same time they send a signal back to retailers and producers.
- PCFs make sense only if they are part of a constant CO_2 reduction strategy of the firms across the value chain. They will complement the instruments of Corporate Carbon Footprints, Carbon Offsetting, and Emissions Trading, which are also evolving in the market.
- PCFs will help to spur innovation processes across the value chain, both technological ones (such as new refrigerators, more efficient engines or less carbon intensive materials) and organizational ones (such as energy round tables, more efficient logistics, or more stakeholder involvement). They can help to translate the given innovation potential into real innovation processes.

Interestingly, it has been the private sector – together with science organizations and environmental NGOs – that has set the pace for PCFs in Europe. Despite the fact that many firms have initially been reluctant to add another label on their products, the international dynamics of the climate and CO_2 label debate has convinced many to move forward. Given the high recent public attention to climate change, it is very tempting for a corporation to demonstrate its pro-active engagement – even if this claim is not really backed by real improvements. This risk of green-washing can only be reduced by transparent procedures, by using scientifically sound and widely accepted methodologies, and by independent certification. These are the criteria that the German PCF Pilot Project does follow. National governments as well as the European Commission are also considering CO_2 labels, and it remains open whether we will see a harmonized, EU wide solution based on a government or based on a private driven label. In any case, PCFs and CO_2 labels provide a very promising tool in powering innovation processes towards the low-carbon economy we urgently need.

Lifestyle Changes, Social Diffusion, and the *Gestaltwandel* of Environmental Issues

If we move away from wind energy and product carbon footprints, we have to embed technological and consumption changes into the wider picture of social change in general. For this purpose, I will come back to the meso-level of lifestyles as group-specific forms of leading and interpreting life, sketched briefly in Section 3.2. My goal is not only to flesh out the somehow ‘skinny’ sketch of lifestyle changes in Germany given in the former sections, my goal is also to correct the notion of social diffusion, based upon the concept of lifestyle dynamics. I will argue that the social diffusion of pro-environmental attitudes and behaviors is not adequately captured by assuming its stepwise entry in social groups other than the initial pioneers, but goes along with changes in the meaning and content of pro-environmental attitudes and behaviors themselves. For these internal changes I would like to use the German expression *Gestaltwandel*.

In order to grasp the meaning of this term, we have to consider changes in the structure of social milieus, as analyzed by the German–French market research institute SinusSociovision (cf. www.sinus-milieus.de). Individual lifestyles are situated in larger social groups – social milieus – which make up the social fabric of societies.[10] Social change not only refers to a changing social landscape, with particular groups emerging, growing or declining as time goes by. It also affects the framing of issues, that is the overall interpretation and evaluation of an issue, as well as the concrete 'tokens' (e.g. activities, associations, things,...) that social groups associate with a particular issue. These changes may comprise fashion, but go far beyond it, as they are linked with social status dynamics and value systems. This also holds for the issue 'environment' (or 'ecology').[11]

Wippermann (2005) has looked at the diffusion of 'ecology' (attitudes, behaviors) across the German society since the early 1980s by applying the Sinus milieu approach. Starting with the 'Alternative Milieu', the 'breeding ground' for the environmental social movement and the Green party during the late 1970s/early 1980s, concern for the environment made its way through the upper segments of the German society, that is the higher status groups. A closer look shows, however, that not all higher status groups are pro-environmental, and not all lower status groups reject the issue.

But this is not my main point. More importantly, the whole social landscape is changing, and a major change being that the 'Alternative Milieu', accounting for about 2% of the German adult population in 1982, has vanished from the map since the early 1990s. It has been in part replaced by the 'Post-Materialists', a modernized follow-up group, aesthetically and politically less visible and radical, but still strongly supporting post-materialist values and life goals. Leading an alternative life, engaging for the environment, or being an ecological avant-garde, is no longer the vision for the members of the post-materialist milieu. Ecology has lost its radical image. With reference to Max Weber, one might speak of a transformation of charisma to everyday life; other scholars have described this process as the 'normalization' of ecology (Brand et al. 1997).

One should notice, however, that – at least according to Sinus – post-materialism is no longer the 'spearhead' of value development in Germany. Since the 1990s, post-materialism has been 'outperformed' by issues like experimentalism, hedonistic orientations, and preferences for value hybrids. Sinus approaches groups situated further 'right', such as 'Modern Performers' or 'Experimentalists', like the value avant-garde of German society, taking the lead in many consumption and leisure trends, as well as in the domain of life-work-balance. For them, environmental

---

[10] Social milieus can be read as a modernized version of social class, as they combine social status with values and everyday cultural practices.

[11] These two terms mean different things. However, in social discourses, the scientific term 'ecology' has been transformed (during the 1970s and early 1980s) into a value based political concept.

issues are far less relevant than for the Post-Materialists. Measured in terms of standard questionnaires designed by environmental sociologists (sounding like 'Yes, I would bring sacrifices for the environment, even if others don't act similarly'), members of this group perform rather badly.

But this does not necessarily mean that the fate of the environment is not important to them. It could also indicate that well-established interview scales become outdated over time, and that we have to face lifestyle changes as part of a sustainability transition. Members of the 'Modern Performer' segment for example expect societies and technologies to be environmentally friendly, and they would like to see governmental regulation and professional systems (including codes of conduct) in place in order to deal with the environmental side of societies' metabolism in a sustainable and professional manner. One might interpret these wishes as an expression of the famous 'attitude–behavior gap', or as another example for the Not-In-My-Backyard (NIMBY) syndrome. I would suggest another reading, however. To put it in a slightly exaggerated manner: Modern Performers display a rational attitude towards social complexity, the division of labor, and the historical evolution of modernity. They reflect adequately that the individual consumer cannot solve collective good problems that require public decision making. In other words: their rejection of a pro-active individual engagement for the environment reflects the difference between the consumer and the citizen – and correctly so. From a social science point of view, it would simply be naïve and overburdening if we would expect individual consumers to behave in ways that were paradigmatically realized by the early Alternatives, for whom ecology has been crucial for their personal and political life, if not an ideology. Articulating expectations about governmental action and professional dealing with environmental problems is both a rational and a historically possible attitude – given the evolution of the issue during the last 30 years.

This is not meant to be an apologia for the 'Modern Performer'. He or she displays consumption and leisure behaviors that 'consume' a lot of resources, and indeed can be improved. My point is a different one: The social diffusion of an issue (an idea, a technology, or a behavior) across society is not congruent with the transfer of an identical 'thing' from one hand to another. Both the subjects and the issue change, especially if the issue is as complex and far reaching as 'the environment' (or better: our interactions with the environment). Today, after the 'normalization' of ecology, there is not only a green environmentalism but also a conservative and a post-modern one in Germany.[12]

[12]For reasons of brevity I cannot illustrate this statement. The rise of organic food in many Western countries, especially in the UK, the US and Germany, would provide another example. The organic movement, with a history dating back to the beginning of the twentieth century, experienced similar rigidities and was supported by similarly marginal groups as renewable energy. Over time, advocacy groups were able to influence national and international (European Union) legislation (labeling, production standards, subsidies etc.), changing the boundary conditions of the system. Growing consumer demand together with citizen based policy reforms have led to intensified growth. Distribution channels have changed accordingly (from farmer markets to supermarkets), price premiums diminish due to economy of scale, the pressure on organic farmers is growing too, and by now many pioneers ask themselves whether this systemic change was really intended – or if it would not be better to go back to the 'pure' origins.

Different social actors select different *aspects* of an issue. Conservatives select issues like nature conservation or water saving, whereas Post-Materialists prefer renewable energy, and Modern Performers choose eco-design and health food. More importantly, all these groups adhere to different *interpretations* of what it means to be 'pro-environmental', and they *frame* the issue differently. Conflicts do no longer arise between environmental and non-environmental attitudes (exclusively), but mainly between different interpretations of what environmentalism means in concrete terms. This has implications for design, for communication, for marketing, and for policy preferences. The green issue is changing its *gestalt*, that is its content, shape, interpretation, and framing. Environmental sociologists should in a way be happy about this change. But it requires their special attention, as well-established instruments and approaches might in turn be found inappropriate.

I would like to make an additional point at the end. Most diffusion models assume homogeneity of the population. In this case, diffusion over time has the usual S-shaped form. If, on the other hand, the population is non-homogeneous, for example due to different network densities, the overall diffusion rate will look differently. In social reality, people differ in terms of their milieu affinity and lifestyles, and these distinctions (Bourdieu 1976) affect the overall speed and social localization of diffusion.

This is an additional argument for more lifestyle-oriented research on environmental consumption. Computer modeling, hardly used in environmental sociology, seems to be a very promising tool for dealing with lifestyle dynamics. Non-linear systems behavior might occur, and for sustainability policies it might be relevant to learn what diffusion pathways might occur, and if there are 'tipping points' (Galdwell 2000) for strategic decisions.

Conclusion

This chapter underlines the necessity for a sustainability transition. Business-as-usual strategies or even incremental (technological) changes will not be adequate by far to reduce the detrimental environmental impacts of modern societies. Catch-up processes and globalization aggravate the situation, and stress the necessity of substantial changes. Many scholars (mostly mainstream economists) and politicians seem to assume that more or less radical changes in the technological sub-structure of modern societies will do the job. Thus no substantial policy changes, no lifestyle changes. The chapter argues against such an assumption. Even radical technological changes will only come about if policies change and if these policies are supported by the citizens and accepted by the consumers.

I propose to give up a narrow view on consumption as an individual process of purchase and use, and instead contextualize consumption, both with regard to the environment and the social reality of modern societies. The concept of lifestyle dynamics is used to provide such a conceptual framework at the social meso-level.

My suggestion is to take the influence of consumers and citizens on technologies and policies seriously into account, as illustrated by the case studies from the

German wind energy sector, the diffusion of green lifestyles over time, and recent initiatives for Product Carbon Footprints. No individual case resembles the other, and so drawing general conclusions from the German case alone would not be a wise move. On the other hand, a closer look at analogous cases in other countries might reveal similarities and differences that become important for mutual learning.

References

Anheier H, Glasius M, Kaldor M (eds) (2001) Global Civil Society 2001. Oxford University Press, UK

Bechberger M, Reiche D (2005) Europe banks on fixed tariffs. New Energy 2 April:14–18

Bechberger M, Reiche D (eds) (2006) Ökologische transformation der Energiewirtschaft Erfolgsbedingungen und Restriktionen. Erich Schmidt Verlag, Berlin

Beck U, Giddens A, Lash S (1999) Reflexive modernization. Polity Press, Cambridge; Stanford University Press, Palo Alto, CA

Bhalla AS, Yao S, Zhang Z (2003) Causes of Inequalities in China, 1952 to 1999. J Int Dev 15:939–955

Botzen WJW, Gowdy JM, van den Bergh JCJM (2008) Cumulative CO_2 emissions: shifting international responsibilities for climate debt. Climate Policy 8:569–576

Bourdieu P (1976) Distinction: a social critique of the judgement of taste. Harvard University Press, Cambridge, MA

Brand K-W, Eder K, Poferl A (1997) Ökologische Kommunikation in Deutschland. Westdeutscher Verlag, Opladen

Buttel FH (2003) Environmental Sociology and the Explanation of Environmental Reform. Organization & Environment, 16(3):306–344

Byzio A, Heine H, Mautz R, Rosenbaum W (2002) Zwischen Solidarhandeln und Marktorientierung. Ökologische Innovation in selbstorganisierten Projekten – autofreies Wohnen, Car Sharing und Windenergienutzung. Soziologisches Forschungsinstitut der Georg-August-Universität, Göttingen

Campbell C (1987) The romantic ethic and the spirit of modern consumerism. Blackwell, Oxford

Trust C (2007) Carbon footprints in the supply chain: the next step for business. The Carbon Trust, London

Carrigan M, Szmigin I, Wright J (2004) Shopping for a better world? An interpretative study of the potential for ethical consumption within the older market. J Consum Mark 21(6):401–417

Christensen P (1997) Different lifestyles and their impact on the environment. Sustain Dev 5:30–35

Cogoy M (1999) The consumer as a social and environmental actor. Ecol Econ 28:385–398

Cohen MJ, Comrov A, Hoffner B (2005) The new politics of consumption: promoting sustainability in the American marketplace. Sustain: Sci Pract Policy 1(1):1–19

Consumers International (1997) A discerning middle class? Sustainable consumption: a preliminary enquiry of sustainable consumption trends in selected countries in the Asia Pacific region. Consumers International. Regional Office for Asia and Pacific

Cowan RS (1983) More work for mother: the ironies of household technology from the open hearth to the microwave. Basic Books, New York

Crocker D, Linden T (eds) (1998) Ethics of consumption: the good life, justice, and global stewardship. Rowman & Littlefield, Lanham, MD

Diekmann A, Preisendörfer P (2003) Green and greenback. The behavioral effects of environmental attitudes in low-cost and high-cost situations. Rational Soc 15(4):441–472

Doubleday R (2004) Institutionalizing non-governmental organization dialogue at Unilever: framing the public as 'consumer-citizens'. Sci Public Policy 31(2):117–126

Duchin F (1998) Structural economics. Measuring change in technology, lifestyles, and the environment. Island Press, Washington DC
Durning A (1992) How much is enough? The consumer society and the future of the Earth. W.W. Norton & Co, New York
Environics (2002) Consumerism: a special report. Environics International Ltd, Toronto
Fine B, Leopold E (1993) The world of consumption. Routledge, London
Fischer-Kowalski M, Amann C (eds) (2001) Societal metabolism and human population. Special issue of population and environment. A J Interdiscip Stud
Fuchs DA, Lorek S (2005) Sustainable consumption governance: a history of promises and failures. J Consum Policy 28:261–288
Galdwell M (2000) The tipping point. How little things can make a big difference. Little, Brown & Co., Boston, FL/New York/London
Giddens A (1986) The constitution of society: outline of the theory of structuration. University of California Press, Berkeley, CA
Guinee JB (ed) (2002) Handbook on life cycle assessment. Kluwer Academic, Dordrecht
Hall PA, Soskice DW (eds) (2001) Varieties of capitalism: the institutional foundations of comparative advantage. Oxford University Press, Oxford
Hernes G (1976) Structural change in social processes. Am J Sociol 38(3):513–46
Hertwich E, Briceno T, Hofstetter P, Inaba A (eds) (2005). Sustainable consumption: the contribution of research. In: Proceedings, NTNU Report 1/2005, Oslo
Huber J (2000) Towards industrial ecology: sustainable development as a concept of ecological modernization. In: Andersen M, Massa I (eds.) Ecological modernization. J Environ Policy Plan, Special Issue 2:269–285
Hughes TP (1983) Networks of power. Electrification in Western society, 1880–1930. John Hopkins University Press, Baltimore, MD
IPCC (Intergovernmental Panel on Climate Change). Climate Change (2007) Synthesis Report. http://www.ipcc.ch/pdf/assessment-report/ar4/syr/ar4_syr.pdf
Jacobsson S, Lauber V (2006) The politics and policy of energy system transformation – explaining the German diffusion of renewable energy technology. Energy Policy 34:256–276
Jackson T (2004) Models of Mammon: a cross-disciplinary survey in pursuit of the 'sustainable consumer', ESRC Sustainable Technologies Programme Working Paper No. 2004/1, Centre for Environmental Strategy, University of Surrey, Guildford, Surrey
Jackson T (2005) Motivating sustainable consumption. A review of evidence on consumer behaviour and behavioural change. A report to the sustainable development research network, Centre for Environmental Strategy. University of Surrey, Guildford, Surrey
Kates RW, Clark WC, Corell R, Hall JM, Jaeger CC, Lowe I, McCarthy JJ, Schellnhuber HJ, Bolin B, Dickson NM et al (2001) Sustainability science. Science 292:641–642
Kemp R, Schot J, Hoogma R (1998) Regime shifts to sustainability through processes of niche formation: the approach of strategic niche management. Technol Anal Strateg Manage 10(2): 175–195
Kemp R, Loorbach D (2003) Governance for Sustainability Through Transition Management. Conference Paper. http://citeseerx.ist.psu.edu/viewdoc/download?doi=10.1.1.10.8115&rep=rep1&type=pdf (date accessed 17 Jan 2008)
Kemp R, Loorbach D, Rotmans J (2007) Transition management as a model for managing processes of co-evolution towards sustainable development. International J Sust Dev & World Ecol 14(1):78–91
Lebel L (2005) Transitions to sustainability in production–consumption systems. J Ind Ecol 9(1–2):11–13
Lüdtke H (1989) Expressive Ungleichheit. Zur Soziologie der Lebensstile. Leske & Budrich, Opladen
Lundvall B-A (ed) (1995) National systems of innovation – towards a theory of innovation and interactive learning. Pinter Publishers, London
Lutzenhiser L, Hackett B (1993) Social stratification and environmental degradation: understanding household CO_2 production. Soc Probl 40(1):50–73
Manoochehri J (2002) Post-rio 'sustainable consumption': establishing coherence and a common platform. Development 45(3):47–53

MGI (McKinsey Global Institute) (2006) From 'made in China' to 'sold in China': the rise of the Chinese urban consumer. McKinsey & Co, Los Angeles, CA
MGI (McKinsey Global Institute) (2007) The 'bird of gold': the rise of India's consumer market. McKinsey & Co, Los Angeles, CA
Müller H-P (1992) Sozialstruktur und Lebensstile. Der neuere theoretische Diskurs über soziale Ungleichheit. Suhrkamp, Frankfurt am Main
Myers N, Kent J (2003) New consumers: the influence of the affluence on the environment. Proc Natl Acad Sci 100(6):4963–4968
National Research Council (NRC) (1999) Our common journey: a transition toward sustainability. National Academy Press, Washington, DC
Porter ME (1990) The competitive advantage of nations. New York: Free Press
Princen T, Maniates M, Conca K (eds) (2002) Confronting consumption. MIT Press, Cambridge, MA
Redclift ME (1996) Wasted: counting the costs of global consumption. New York: Free Press
Reusswig F (1994) Lebensstile und ökologie. Verlag für Interkulturelle Kommunikation, Frankfurt am Main
Reusswig F, Gerlinger K, Edenhofer O (2004) Lebensstile und globaler Energieverbrauch. Analyse und Strategieansätze zu einer nachhaltigen Energiestruktur. Potsdam Institute for Climate Impact Research, PIK-Report No. 90. http://www.pik-potsdam.de/pik_web/publications/pik_reports/reports/pr.90/pr90.pdf (Date accessed 18 Jan 2009)
Robison R, Goodman DSG (eds) (1996) The new rich in Asia. Mobile phones, McDonald's and middle-class revolution. Routledge, London/New York
Røpke I (1999) The dynamics of willingness to consume. Ecol Econ 28:399–420
Røpke I (2005) Consumption in ecological economics. Internet encyclopaedia of ecological economics. http://www.ecoeco.org/pdf/consumption/in ee.pdf
Rotmans J, Kemp R, Asselt Mv (2001) More evolution than revolution: transition management in public policy. Foresight – The journal of future studies, strategic thinking and policy 3(1) February 2001:15–31
Schaefer A, Crane A (2005) Addressing sustainability and consumption. J Macromark 25(1):76–92
Schellnhuber H-J et al (2006) Avoiding dangerous climate change. Cambridge University Press, Cambridge
Shove E, Warde A (1998) Inconspicuous consumption: the sociology of consumption and the environment. Department of Sociology, Lancaster University. http://www.comp.lancs.ac.uk/sociology/papers/Shove-Warde-Inconspicuous-Consumption.pdf
Spaargaren G (2003) Sustainable consumption: a theoretical and environmental policy perspective. Soc Natl Resour 16(8):687–701
Sridharan E (2004) The growth and sectoral composition of India's middle class: its impact on the politics of economic liberalization. India Rev 3(4):405–428
Stern PC, Dietz T, Ruttan VW, Socolow RH, Sweeney JL (eds) (1997) Environmentally significant consumption: research directions. National Academy, Washington, DC
Stevenson N (2002) Consumer culture, ecology and the possibility of cosmopolitan citizenship. Consumpt, Mark Cult 5(4):305–319
Stier B (1999) Staat und Strom – Die politische Steuerung des Elektrizitätssystems in Deutschland 1890–1950. Verlag Regionalkultur (Ubstadt.Weiher)
Tomlinson M (2003) Lifestyle and social class. Eur Sociolog Rev 19(1):97–111
Uusitalo L (1986) Environmental impacts of consumption patterns. Gower, Aldershot
Van Wessel M (2004) Talking about consumption. How an Indian middle class dissociates from middle-class life. Cult Dyn 16(1):93–114
Walker G, King D (2008) The hot topic. What we can do about global warming. Harvest Harcourt, Orlando, FL
Weber C, Perrels A (2000) Modelling lifestyle effects on energy demand and related emissions. Energy Policy 28:549–566
Weingart P, Engels A, Pansegrau P (2000) Risks of communication: discourses on climate change in science, politics, and the mass media. Public Underst Sci 9:261–283

Wippermann C (2005) Die soziokulturelle Karriere des Themas ‚Ökologie': Eine kurze Historie vor dem Hintergrund der Sinus-Lebensweltforschung. Heidelberg, Sinus Sociovision. http://www.sinus-sociovision.de/Download/karriere_oekologie.pdf

York R, Rosa EA, Dietz T (2004) The ecological footprint intensity of national economies. J Ind Ecol 8(4):139–154

Zukin S, Maguire JS (2004) Consumers and consumption. Annu Rev Sociol 30:173–197

Chapter 4
Emerging Challenges of Consumer Activism: Protecting Consumers and Advocating Sustainable Consumption in Developing Countries

Uchita de Zoysa

Introduction

Mahatma Gandhi once said that the difference between what we do and what we are capable of doing would suffice to solve most of the world's problems, and that you must be the change you wish to see in the world.

This generally is the case for consumer activism too.

Consumer activism, according to the International Organization for Consumer Unions, is intended to eliminate the frustrations of consumers regarding goods and services. The mission of consumer movements is to redress fundamental imbalances in society for the consumer's benefit and to make society more responsive to the consumer's needs and interests.

The debate on sustainable consumption between developed and developing countries is far from settled. Therefore, this chapter provides a developing country perspective on sustainable consumption that may help consumer activists formulate their visions, strategies and action plans.

Consumer activism in developing countries should be analyzed, appreciated or condemned in terms of what activism is meant to be in the context of a globalised world that is dominated by western consumerist lifestyles, political systems and trading networks.

Consumer Activism and Consumer Protection in Developing Countries

Consumer activism in developing countries needs to be understood in the context in which the consumer behaves and is made to behave.

In countries where corruption, exploitation, inequity and injustice are deep-rooted, consumer activism can be a dangerous vocation. Even with the rise of consumer

U. de Zoysa
National Advisory Committee on Climate Change, Sri Lanka
e-mail: uchita@sltnet.lk

L. Lebel et al. (eds.), *Sustainable Production Consumption Systems: Knowledge, Engagement and Practice*, DOI 10.1007/978-90-481-3090-0_4,

protection legislation, the establishment of consumer protection acts and the presence of consumer authorities, the lack of political will for enforcement makes the consumer movement weak and the consumer uncertain.

Protecting and Helping Consumers

'Consumer protection' is a form of government regulation designed to safeguard the interests of consumers. It is linked to the idea of consumer rights which help consumers make better choices in the marketplace. However, in developing countries, ineffectiveness in the enforcement of the regulations takes back the rights of the consumer and hinders empowerment.

Thus, consumer activism is conducted under trying conditions. It often lacks institutional and financial resources as well as support from regulators and authorities. It operates under conditions where a large majority of consumers live in poverty and lack knowledge of their human and consumer rights. Because the legal system is distant from consumers, redress by enforcement agencies is not trusted by the poor. Capital-rich corporate advertisers can manipulate consumer thinking and behavioral patterns. Their power goes unchallenged due to a lack of the financial resources required to counter false product claims. Low prices which remain the most influential purchasing criteria in poverty ridden societies allow flooding of markets by inferior and counterfeit goods (Hamdan 2000).

A former president of Consumer International, Dato Dr. Anwar Fazal (2001), commented "It is often said that consumers are apathetic and that's their problem. I believe consumers only appear apathetic and this is so because they are often frightened, they often have no confidence in the capabilities of the authorities to deal with their problems, the costs of complaining are too high, and the act often 'frowned' upon."

Under these circumstances, consumer organizations in developing countries tend to react to events such as unfair pricing, price hikes of consumer items, shortages of essential goods, substandard quality of goods, exploitative trade practices, misleading advertising, corporate exploits and market dominations. Thus, consumer organizations are perceived as advocacy groups or monitors of corporate accountability that seek to protect people from corporate abuse and malpractice. They normally operate via protests, campaigns or lobbies and also provide consumers with useful information.

The more organized and internationally networked consumer protection organizations operate on a set of rights of consumers which includes access to basic goods and services, fair prices, safety information, representation, redress, consumer education and healthy environment. The organizations believe that their main responsibilities are critical awareness, action, social responsibility, ecological responsibility and solidarity. However, although a few headstrong activists with a social conscience keep up the fight against corporate exploitation and lack of regulatory response, in fact consumer protection in civil society has now become a professional affair and a select vocation for many involved. This change of character of consumer protection organizations from activism to professionalism has raised questions about the legitimacy of some consumer movements.

The other side of the story in developing countries is the insufficient knowledge of consumers to make sustainable consumption choices. Although consumer purchasing power is limited, the inability to make the right choice also stems from lack of awareness, education and knowledge. The main reasons for the low levels of consumer awareness are weak consumer movements, insufficient mainstream education, and inadequate public awareness programs.

In Asia, cultural awareness and education drive a general consciousness and willingness to act pro-environmentally. Both are much stronger tools than law and enforcement. However, as poverty is associated with pollution and environmental destruction at the community level, the harmful consequences of wrongful consumer behavior and the valuable benefits of positive consumer behavior are generally not observed by developing countries and their urban consumers in particular.

Very few developing countries have a consumer activism strong enough to challenge violations and provide redress. India is one such country where the consumer movement is mature and active due to having established legal and legislative systems. Pushpa Girimaji (1999) writes in her book 'Consumer Rights for Everyone' (1999) that if there is one date that will forever be etched in the memory of those who are committed to the empowerment of consumers in India, it is 24 December 1986, the date on which the Consumer Protection Act was passed by India's parliament.

Emergence of Consumer Activism

Stemming from the civil rights and social justice movements in the USA, the consumer movement began there with the objective of protecting the rights of the consumer. The basic rights of American consumers, as set forth by U.S. President John F. Kennedy in his 1962 message to Congress, were (1) the right to safety; (2) the right to be informed; (3) the right to choose; and (4) the right to be heard. Industrialization brought to other countries the same kinds of consumer problems it had caused in the U.S., and the responses to these problems has been similar.

Consumer activism then turned international with the birth of Consumers International (CI), a nongovernmental organization (NGO) that represents consumer groups and agencies all over the world. CI has a membership of over 220 organizations in 115 countries. Its goal is to promote a fairer society by defending the rights of all consumers, by increasing the number and scope of consumer organizations, and by campaigning for 'consumer friendly' policies and laws. It was established in 1960 as the International Organization of Consumers Unions (IOCU) by national consumer organizations. The original members recognized that they could build upon their individual strengths by working across national borders.

Consumers International claims to be the voice of the international consumer movement on issues such as product and food standards, health and patients' rights, and regulation of international trade and public utilities.

The emergence of consumer movements in developing countries was reflected throughout Latin America. In the 1990s consumer activism rose to become one of the most visible new forms of political mobilization these countries had seen since the advent of movements against military rule. As traditional class-based organizations declined during this era of globalization, innovative political movements together with civil society leaders and grassroots organizations used consumer protection issues to influence public policy.

Sybil Rhodes (2005) argues that privatization and deregulation of the telecommunications sectors in Chile, Argentina, and Brazil in the 1980s and 1990s provoked the rise of new consumer protest movements in Latin America. Hasty privatization of state-owned telephone companies led to short-term economic windfalls, but long-term instability, for multinational corporations due to consumer boycotts or the threat thereof. Eventually these governments implemented consumer-friendly regulation as a belated form of damage control. In contrast, governments that privatized through more gradual, democratic processes were able to make credible commitments to their citizens as well as to their multinational investors by including regulatory regimes with built-in consumer protection mechanisms. Sybil Rhodes argues that the relationship between privatization and consumer mobilization occurred because it was embedded in the political context of democracy. She believes that the new consumer consciousness was forged by Latin American politicians, lawyers and activists whose organization of consumer movements contributed to changing the substance of democratic politics in the region from a class base to a consumer base.

Anti-privatization campaigns brought out the activists in consumer organizations all across the developing world during the 1990s. Activists in Latin America as well as in many other parts of the world such as South East Asia received support from global civil society. During the latter half of the 1990's, transnational Latin American networks for consumer advocacy developed and then received technical and financial aid from abroad. International activism had a multiplier effect on local consumer initiatives which resulted in the creation of new organizations to represent consumer issues.

In the meantime, consumers in developing countries face important threats every day. These are problems such as poverty and hunger, global warming and climate change, depletion of natural resources and biodiversity, crash in value of capital resources and market fluctuations, etc. Because of these issues now affecting the consumer, the role of consumer activism in the developing world needs to be examined beyond consumer protection. This need has reached the more internationalized consumer organizations that seem to be expanding into issues that are not traditionally scripted in the framework of the market system. These organizations have thus extended their efforts to consumers as people with emotions, needs, desires, cultures and environments in which sustainable consumption occupies a central role.

Consumer Activism and Sustainable Consumption in Developing Countries

The concepts of sustainable consumption and consumer activism are not as convergent in the developing countries as they are in Europe. Until sustainable consumption is fully comprehended and accepted, consumer activism in the developing world will be based on a rights agenda that places price, quality and quantity at the top of the priority list.

Advocating Sustainable Consumption

Developing countries rarely find poverty eradication adequately addressed within the global regulatory processes for sustainable consumption. European-dominated, efficiency-based sustainable consumption agendas have marginalized the priorities for sufficiency of developing countries. Thus, the United Nations process to establish a 10-year framework of programs for sustainable consumption and production (UN 2003) has not gained much importance in the national policy circuits or the civil society action lists. Sustainable consumption is still an environmental consideration for most people in these parts of the world, where consumers continue to struggle for survival.

In developing countries, sustainable consumption issues cannot be separated from basic development objectives such as poverty alleviation, access to basic services, gender equality, and environmental protection. The goal of attaining such basic goods and services as food, clean water, sanitation and shelter remains largely a dream for many consumers. Lifestyles, human behavior and consumption patterns affect these development objectives.

In many ways, the continued mistrust between business sectors and consumer protection organizations keeps consumer activism alive, even though such mistrust may not foster the best way of finding solutions. Inequity in the global trading system that allows trans-national corporations to exploit consumers in developing countries has been a cornerstone of the battles fought by consumer movements. The campaigns against globalization measures and structural adjustments have been vociferous, but lack the money and power to sustain the rebellion that could bring about radical changes to global systems. The emergence of the Group of 77, a forum of developing countries, against the western economic powers and the trade regimes favoring multinational corporations, is a boost for consumer activism in the southern hemisphere and developing countries.

Lifestyles in developing countries are no longer as conservative and environmentally friendly as they used to be. Growing economies, urbanization, and globalization have underpinned the rise of 'happiness-through-consumption' in developing countries. More and more people are joining the ranks of the 'consumer classes'. For example, the 2004 State of the World Report says that, in China alone, the consumer class of 2002 consisted of 240 million people. Research by the McKinsey Global Institute suggests that by 2025 China will be one of the world's largest consumer markets with households spending 22.6 trillion RNB ($2.5 trillion) every year, which is as much as Japanese households spend today.

Another growing reality is urbanization. Approximately half the world's population now lives in cities and towns. Furthermore, in 2005, one out of three urban dwellers (approximately 1 billion people) lived in slum conditions. According to the Asian Review on Sustainable Consumption (Zoysa 2007), growing economies, urbanization, and globalization are linked to the rise of 'happiness-through-consumption' in developing countries. In Asia, the ambition of China and India is to become developed countries within the next 10 years; Thailand wants to be the kitchen of the world and the fashion hub of Asia. Even though cities such as Beijing, Jakarta, New Delhi, Bangkok, Manila, and Dhaka are highly populated, congested and polluted, their inhabitants want to be modern, sophisticated, and trendy. Clearly, Asia has already embarked on the path of rapid economic growth to raise the living standards of its citizens. Asia holds a majority of the earth's consumers, is home to 70% of the global poor, owns a large percentage of the world's natural resources, and boasts 12 cities with the highest levels of particulate pollution: all these factors pose a serious obstacle to sustainable consumption in Asia.

The key to sustainable consumption in developing countries lies in their traditional and cultural background of sustainable livelihoods. Religious observances, traditional beliefs, and close-to-nature lifestyles are still practiced by rural populations and respected by city adults. While the winds of 'happiness-through-consumption' blow across these nations, low consumption lifestyles also provide politicians and regulators with opportunities to introduce sustainable consumption choices to their constituencies. However, the wellbeing of these people will depend on a new socio-economic regime which provides more equitable distribution of resources and wider opportunities for consumption. This is confirmed by the 'Asian Review on Sustainable Consumption', (Zoysa 2007) (see Appendix for more details), which identified 'Wellbeing' and 'Happiness' as the main objectives of sustainable consumption in Asia, but which also recognized 'Better Quality of Life' as the primary goal of its people.

Fight for a Share of Consumption in an Imbalanced Global Trading System

Consumer activism aims to provide a better quality of goods and services to consumers and to thereby assure a better quality of life for consumers. In developing countries, the struggle for a better quality of life is clearly made more difficult by over-consumption in developed countries. While world gross domestic product was US$48.2 trillion in 2006, the wealthiest 20% of the world accounted for 76.6% of total private consumption, the poorest fifth had just 1.5% and the middle 60%'s share was 21.9% (for a world population of approximately 6.5 billion). The poorest

10% accounted for 0.5%, the wealthiest 10% accounted for 59% of all consumption, an unsustainable imbalance.

According to the World Bank and other sources, the world's wealthiest countries (approximately 1 billion people) accounted for $36.6 trillion dollars or 76% of GDP (Gross Domestic Product), low income countries (2.4 billion people) contributed just $1.6 trillion (3.3%) and middle income countries (3 billion people) made up the rest of GDP at just over $10 trillion (20.7%). The GDP of the 41 most heavily indebted poor countries (567 million people) is less than the combined wealth of the world's seven richest people. Whereas the world's poorest countries accounted for just 2.4% of world exports, the total wealth of the top 8.3 million people around the world rose 8.2% to $30.8 trillion in 2004, giving them control of nearly a quarter of the world's financial assets; in other words, about 0.13% of the world's population controlled 25% of the world's financial assets in 2004. Furthermore, for every $1 in aid it receives, a developing country spends over $25 on debt repayment.

While developing countries suffer in an unfair global trading regime, the labor and services of these regions are exploited by multinational corporations and by inter-regional trading mechanisms. Low wages for Asian labor and dumping of inferior products into Asian markets continue on a large scale and threaten the emergence of sustainable lifestyles (for examples see Chapter 7 this volume). The United Nations Development Programme (UNDP) states that recent trends in trade will damage human development, with serious implications for the Least Developed Countries (LDCs) and also for the poorest people in the larger and more dynamic economies, where trade is accompanied by inadequate job creation and rising inequality. In fact, Asian LDCs suffer in international trade because they are minor players who make a small contribution to the global market and thus cannot influence international prices. They become 'price takers' and have to accept what they are offered. If other producers increase their output, thus pushing down international prices, the LDCs' already-low profit margins will be wiped out. In the short term, this can deliver a short, sharp economic shock or in the longer term push the LDCs into a 'poverty trap'.

Martin Khor of the Third World Network, in his book 'Globalization and the South' (2000) writes that the developing countries' weakness stems from their lack of bargaining and negotiating strength. Because they are heavily indebted and dependent on bilateral aid, developing countries do not even have the capacity to negotiate loan conditionality. He argues that developing countries are not as well organized as developed countries to deal with rapid developments in the global economy because, in developing countries, the links between intellectual sectors, NGOs and governments are often too weak to deal with change.

In the report 'From Cancun to Sao Paulo: The Role of Civil Society in the International Trading Systems', the rapporteur writes that many NGOs, and indeed governments of developing countries, do not know the strengths (and limitations) of intergovernmental institutions (Diana 2004). This was written about the specific issues of capacity building and of generating an international debate on issues of trade and development. In effect, according to the rapporteur, developing countries have weak domestic policy processes relating to issues of trade and development. They are also weak at forging alliances with interests in developed countries that could help shape policies in those countries to the advantage of the developing country. This governmental weakness is mirrored by that of NGOs in those situations.

The low share of consumption of developing countries can thus be attributed to their weak bargaining power in global markets. In turn, lack of bargaining power has led to exploitation by market forces as well as victimization by dominant global trading systems.

Consumer Activism and Sustaining Consumers in Developing Countries

The gradual spread of privatization across the world, the structural adjustment programs of the international banks and the acceptance by governments of the private sector as the engine of economic growth challenge consumer movements. The diminished political support obviously also makes it hard to keep the momentum of the movements alive. Thus, consumer organizations need to adjust to the new realities of a globalised world order.

Fighting for Equity, Fighting Against Poverty

Poverty clearly impedes the wellbeing of consumer-citizens in developing countries. Dr. Mohammad Yunus, founder of the Grameen Bank in Bangladesh sees in poverty the absence of all human rights. In his Nobel Prize acceptance speech of 2006, Dr. Yunus explained: “I believe that we can create a poverty-free world because poverty is not created by poor people. It has been created and sustained by the economic and social system that we have designed for ourselves; the institutions and concepts that make up that system; the policies that we pursue. Poverty is created because we built our theoretical framework on assumptions which underestimates human capacity, by designing concepts, which are too narrow (such as concept of business, credit-worthiness, entrepreneurship, employment) or developing institutions, which remain half-done (such as financial institutions, where poor are left out). Poverty is caused by the failure at the conceptual level, rather than any lack of capability on the part of people” (Yunus 2006).

Consumers in developing countries come from the pool of global poor – that is almost half the world or over 3 billion people – who live on less than US$2.50 a day. More than 80% of the world’s population lives in countries where income differentials are widening. The poorest 40% of the world’s population accounts for 5% of global income. New poverty data released in August 2008 by the World

Bank estimate that 1.4 billion people in the developing world (one in four) lived on less than $1.25 a day (new poverty line) in 2005, down from 1.9 billion (one in two) in 1981.

Halfway to the 2015 deadline for the Millennium Development Goals (MDGs), millions of people still live in poverty and hunger in sub-Saharan Africa and Southern Asia. If we cannot overcome poverty and hunger, all other development goals within the MDG will also fail.

The issue of food security in Africa provides shocking statistics. Eight hundred and forty million people were undernourished during the period 1998–2000. Chronic food insecurity affected about 28% of the African population, that is 200 million people suffered from malnutrition. Acute food insecurity affected 38 million people in Africa in 2003. Every day, 24,000 people die of hunger. According to the FAO, of the 39 countries that face food emergencies worldwide, 25 are in Africa. About 180 million children under the age of five were underweight in 2000; 30% of these children lived in Africa (Rutivi 2008).

However, consumption issues such as hunger and poverty cannot be left to consumer protection organizations. In this regard, social justice and community empowerment organizations play a key role in developing countries to secure consumer rights. Thanks to this sector, micro-credit has spread to every continent and benefited over 100 million families during the past 30 years. It now provides poor consumers with credit and thus ensures access to basics human needs and goods.

In this respect, the Grameen Bank of Bangladesh is legendary for creating the concept of micro-credit and thus making poor peoples' lives better across developing countries. The Grameen Bank Project was born in Jobra, Bangladesh, in 1976, and was transformed into a formal bank under a special law passed for its creation in 1983. The bank's objective is to bring financial services to the poor, particularly women, and thus to help the poor fight poverty, to stay solvent, and to be financially sound. Grameen Bank raises the status of poor women in their families by giving them ownership of assets. Grameen believes that all human beings, including the poorest, are endowed with endless potential. Borrowers at Grameen Bank own 94% of the equity in the bank. The remaining 6% is owned by the government. 'Grameen Credit' promotes credit as a human right. It rejects conventional banking that labels the poor as "not creditworthy". This bank belongs to the people; so it goes to the people.

Consumer organizations, unlike the Grameen Bank, have shown little urgency to tackle poverty compared to other movements from social empowerment, social justice and even environmental action. The better-off consumer classes within developing country societies place the poor classes as the majority segment and accept that the poor are the most vulnerable to exploitation and the most in need of redress and protection. The consumer movement must realize that the future of sustainable consumption lies in the well-being of over half of humanity which consumes little or has few consumption opportunities. If equity in consumption is to be adequately addressed, then consumption opportunities must be provided for the majority of the consumers of the world.

Ensuring a New Economic Order

The concept of perpetual growth is now widely challenged and the collapse in 2008 starting with the Wall Street-based economy in the USA as well as major financial institutions across Europe has sent a shock wave across global economies and markets. Economic globalization and trade liberalization which were the key vehicles for growth-related development may no longer be adequate to provide the prosperity and economic sovereignty for the emerging economies of developing countries.

In 1972, the Club of Rome declared in The Limits to Growth: (1) if the present growth trends in world population, industrialization, pollution, food production, and resource depletion continue unchanged, the limits to growth on this planet will be reached sometime within the next 100 years. The most probable result will be a sudden and uncontrollable decline in both population and industrial capacity; (2) it is possible to alter these growth trends and to establish a condition of ecological and economic stability that is sustainable far into the future. The state of global equilibrium could be designed so that the basic material needs of each person on earth are satisfied and each person has an equal opportunity to realize his or her individual human potential; (3) if the world's people decide to strive for this second outcome rather than the first, the sooner they begin working to attain it, the greater will be their chances of success.

Twenty years later, in the book 'Beyond the Limits to Growth', the same team of researchers, Donella H. Meadows, Dennis L. Meadows, and Jørgen Randers, concluded: (1) Human use of many essential resources and generation of many kinds of pollutants have already surpassed rates that are physically sustainable. Without significant reductions in material and energy flows, there will be in the coming decades an uncontrolled decline in per capita food output, energy use, and industrial production; (2) this decline is not inevitable. To avoid it two changes are necessary. The first is a comprehensive revision of policies and practices that perpetuate growth in material consumption and in population. The second is a rapid, drastic increase in the efficiency with which materials and energy are used; (3) a sustainable society is still technically and economically possible. It could be much more desirable than a society that tries to solve its problems by constant expansion. The transition to a sustainable society requires a careful balance between long-term and short-term goals and an emphasis on sufficiency, equity, and quality of life rather than on quantity of output. It requires more than productivity and more than technology; it also requires maturity, compassion, and wisdom (Meadows et al. 1992).

New analysis indicates that global economic growth has been an extremely inefficient way of achieving poverty reduction and is becoming even less effective (Hamilton 2003). Between 1990 and 2001, for every US$100 worth of growth in the world's income per person, just $0.60 found its target and contributed to reducing poverty below the $1-a-day line. To achieve every single $1 of poverty reduction therefore requires $166 of additional global production and consumption, with all its associated environmental impacts. This approach is both economically and ecologically inefficient. Thus, if global growth remains the principal economic strategy, the objectives of poverty reduction and environmental sustainability will be highly difficult to reconcile. The scale of growth which this model demands would generate unsupportable environmental costs; and the costs would fall disproportionately, and counter-productively, on the poorest people the growth is meant to benefit.

In the recent history of civil action, the more ecologically oriented civil society movements and thinkers have advocated limits to growth and even no-growth economies. The anti-globalization and anti-privatization movements, as well as the organizations fighting unjust global trading regimes and Northern-favored WTO policies, have also challenged the current economic globalization and trade liberalization policies. Martin Khor (2000) explains that the most important aspects of globalization are the breaking down of national economic barriers, the expansion of international trade, finance and production, and the growing power of transnational corporations and international financial institutions in these activities. Although increased trade and investments may be focused in a few countries, the effects of an increase may nevertheless be felt in almost all countries. For example, a low income country may account for only a miniscule part of world trade, but reduced demand or lower prices for its export commodities can have a major economic and social effect on that country. Martin Khor writes that there is increasing disillusionment about globalization among many policy-makers, research academics and NGOs in the South and he maintains that environmental, social and cultural problems have been made worse by the workings of the global free-market economy.

To counter such world trade inequality and exploitation, the international fair-trade movement launched a major consumer initiative. The campaign for fair trade in the coffee industry is a highpoint of consumer activism that benefits developing countries (Deborah 2002). Coffee is the world's second most valuable commodity after petroleum and is a significant source of foreign exchange for many Latin American and African countries. It was traditionally grown as a colonial crop, planted and harvested by serfs or wage laborers on large plantations, and then shipped to imperial masters.

Farmers, many of them indigenous peoples, now grow much of the world's coffee beans on plots of less than 10 acres. The prices farmers receive are often less than the costs of production, which pushes them into an endless cycle of poverty and debt. All over Latin America, farmers are forced to sell the future rights to their harvest to exploitative middlemen in exchange for the credit required to pay for basic necessities. However, some coffee is also grown on large plantations worked by landless day laborers with low rates of unionization and under extremely poor working conditions. Although, these coffee workers are denied basic labor rights worldwide, efforts to develop an industry-wide Code of Conduct are underway.

The fair trade movement is growing among coffee consumers, especially in the United States, to demand social justice for coffee workers and farmers. Consumer activists have put pressure on big coffee retailers to buy directly from farmer-cooperatives and to pay a price for their coffee that represents a living wage. The movement demands that corporations not only makeover their images as socially responsible businesses, but also account to all their stakeholders; these include consumers, citizens, and, most importantly, the people who produce the goods that generate their revenue. But the movement also encourages Americans to be ethical consumers. Importantly, it demonstrates that when we act not only as consumers, but also as citizens, we can force large corporations to recognize basic principles of human rights and fairness. Because of the fair trade movement, dozens of companies now offer millions of consumers a choice: coffee produced under sweatshop conditions or coffee based on principles of fair trade (for further elaboration on the coffee case see Chapter 10 this volume).

Fair trade movements, anti-globalization campaigns, consumer-rights groups, all consumer activists should now join ecological groups to argue that economic growth has not solved and will not solve the equity, poverty and wellbeing issues of many consumers in the poor classes. The debate needs to be taken beyond distribution of wealth and welfare of citizens. It must essentially move towards a new economic order that ensures equitable consumption opportunities for all within the carrying capacity of the earth.

Conserving the Environment Under a Changing Climate

The right to clean air, water, and space, and the access to natural resources have been the battleground of environmental movements for a very long time. As environmental problems constrained 'happiness-through-consumption' and globalization issues affected environmental activism, the consumer and the environmental movements joined to embrace sustainability issues during recent decades. Topics such as green products, organic farming, genetically modified food, intellectual property rights, food security, toxic contamination, water privatization, waste dumping, and climate change, as well as the influence of an urbanized and westernized consumerist lifestyle have encouraged the consumer movement to see a closer relationship between consumer protection and environmental activism.

In developing countries, sustainable consumption has become a cause for the consumer movements with the involvement of the environmental movements. Although the movements are seen as two separate lobbies, environmental and consumer activists try to find common ground to stop the over consumption of resources by western industrial forces and to protect the consumption opportunities of consumers (UNEP 2005).

For example, Martin Khor (2002) describes the relationships between intellectual property rights (IPR), environmental biodiversity, human development and sustainability. These four topics bring together consumer and environmental activists in a single arena of intergovernmental debates, including the World Trade Organization (WTO), the Convention on Biological Diversity (CBD), the Food and Agriculture Organization (FAO).

Campaigns against genetically modified organisms (GMO) in developing countries also demonstrate joint activism on behalf of food security and protection of biodiversity. Writing about genetically modified crops, Suman Sahai (2003) warns that developing countries must examine gene technology carefully to see what is of relevance to their needs. Because these countries are so rich in biodiversity, they must reflect carefully on what kind of GM crops they can safely use and they must make choices in the best interest of their farmers and consumers. The book also recommends that civil society organizations (CSO) should inform themselves about the issues, legislation and policies in place in their countries for agriculture biotechnology and GM crops. There should be a high level of technical competence, accountability, transparency and public participation in the regulatory system set in place for GM technology. CSOs should be in touch with the scientific community, should keep up to date with new information and should facilitate consultation on a national agenda for GMO.

Climate change is another underlying consequence of unsustainable consumption. A joint international study by Consumers International (CI) and AccountAbility states: "Consumers International believes it is essential and urgent that governments and the business sector everywhere commit to implementing policies that guide us onto a sustainable course. It is equally essential that consumers everywhere are empowered and encouraged to act in relation to climate change. And consumer organizations around the world want to play their part. In a recent survey of CI member organizations in 115 countries, sustainable consumption emerged as the number one international issue" (AccountAbility & CI 2007). This report which is conducted through surveys with US and UK consumers says that civil society organizations are beginning to make the crucial links between global warming, poverty and sustainable development, but struggle to fit climate change into their existing schemas of perpetrators and victims, donors and beneficiaries, global problems and activist solutions.

Climate change in developing countries is more about adaptation and less about mitigation, the top priority and the responsibility of the developed world. Consumer activism in developing countries will therefore need to lead back to issues of well-being while it strikes for consumption increases with a low environmental impact through environmentally friendlier development processes. Consumer movements in emerging countries like China, India, Brazil and South Africa will need to intensify the debate on developing sustainable consumption lifestyles and will have to bring autonomous arguments independent from the influences and guidelines of western consumer organizations.

Consumption has deep roots in the socio-cultural context and therefore can reinforce or safeguard other positive societal values. Issues of food sovereignty and food security are intrinsically connected with careful handling of the soil, the water and biological diversity. By changing what we consume and how we provide products and services we can ensure that the environment is protected. Therefore consumer policies can set guidelines for production and standards for information with respect to environmental factors (Bentley et al. 2002). National consumer organizations can then monitor progress and encourage consumers and industry to act for a more sustainable lifestyle.

Emerging Goals for Consumer Activists in Developing Countries

The former President of India A.P.J. Abdul Kalam states (1998): "A reasonable lifespan, a rewarding occupation, some basic comforts and a good health care system is the dream of a developed India". As well, all people should enjoy a secure present and look forward to a better future. The greatest challenge for consumer activism in developing countries will be to ensure 'better quality of life for all' as consumers search for 'wellbeing' and 'happiness'. These two will be the principal tests for sustainable consumption in Asia, Africa and Latin America.

To this end, consumer movements will need to increase consumer awareness through educational and promotional programmes. They will have to be more sensitive to issues such as waste management, product certification and testing, and implementation of sustainable practices of consumption. Consumer organizations will be obliged to carry out consumer behavior research and then to cultivate more psychological and spiritual development amongst consumers.

It will also be important to strengthen enforcement capacity as well as legislation on consumer protection in developing countries. Effective economic incentives to promote sustainable production through efficient and appropriate technologies will help promote Corporate Social Responsibility (CSR) amongst business and industry.

Key goals for consumer activism in the promotion of sustainable consumption could be to press for the development of sufficiency economies, the alleviation of poverty, the provision of food security and safety, the assurance of health and nutrition, the development of environmentally sound products and services, the enforcement of good governance, and the creation of an informed society.

Former Executive Director of the United Nations Environment Program Klaus Töpfer has said "Consumers are increasingly interested in the world that lies behind the product they buy. Apart from price and quality, they want to know how, where and whom has produced the product. This increasing awareness about environmental and social issues is a sign of hope. Governments and industry must build on that". While Töpfer's statement mostly reflects emerging consumer attitudes in Europe and North America, the large and growing middle class in the developing countries is also embracing the concept he describes. However, this transition cannot fully rely on business and government. Consumer activism needs to be pushed towards a more sustainable consumption orientation if truly sustainable consumer lifestyles are to be realized in the developing countries and the world.

The Asian Review on Sustainable Consumption (see Appendix) has argued that sustainable consumption needs to cater for the interests of half of the world's population that are in poverty and also address the over consumption issues of the developed countries. It emphasized that equity in consumption is a major challenge for an international community that seeks to regulate unsustainable consumption patterns. If the Sustainable Consumption agenda cannot address the basic requirement of most humans on earth such as food, clothing and shelter and cannot understand that issues such as food security, fair trade and good governance are intrinsically linked to creating sustainable consumption, then the process will naturally fail the people's aspiration for a sustainable world.

In this regard, the confident words of the authors of 'Beyond the Limits to Growth' should encourage consumer activists in developing countries: "We think that a better world is possible, and that the acceptance of physical limits is the first step toward getting there. We see 'easing down' from unsustainability not as a sacrifice, but as an opportunity to stop battering against the earth's limits and to start transcending self-imposed and unnecessary limits in human institutions, mindsets, beliefs, and ethics."

Appendix

United Nations Guidelines on Consumer Protection and the Asian Review on Sustainable Consumption

The main lobby tool on sustainable consumption has been '*The United Nations Guidelines on Consumer Protection*'. Unanimously adopted by the UN General assembly on 10th April 1985, it has provided to the organized international consumer groups a platform to make their demands more official. These guidelines were developed to serve as general principles for the protection of consumers, especially those in the developing countries. In 1999, the guidelines were expanded to include elements on sustainable consumption. The expanded guidelines were an important opportunity to connect environmental protection and sustainable development and to strengthen the linkage between consumer interests and sustainable consumption. The guidelines encompass such issues as providing information, conducting consumer research, testing products, promoting recycling and sustainable government practices, encouraging life-cycle thinking and eco-products, and strengthening regulatory mechanisms.

In June 2004, the United Nations Environment Program (UNEP) and Consumers International (CI) released the second edition of the study "*Tracking Progress: Implementing Sustainable Consumption Policies*". This European Review summarized available information materials on policies, tools and other instruments adopted in support of the UN consumer guidelines, especially on sustainable consumption practices, in each of the six EU countries. The results claimed that 'General awareness of the sustainable consumption section of the UN Guidelines was low in both OECD and non-OECD countries. About two-thirds (70%) of OECD countries were aware of their existence, compared with just over half (55%) of non-OECD countries'. In all eight sustainable consumption-related policy areas measured, non-OECD countries have achieved comprehensively lower implementation rates than OECD countries.

In the wake of these survey findings, a project called 'SC. Asia: Capacity Building for Implementation of UN Guidelines on Consumer Protection (sustainable consumption) in Asia' was a landmark process in creating awareness and sensitivity for sustainable consumption in the Asian Region. It was the first such project in the developing country regions. The Asian Review was to undertake a review of existing Government initiatives promoting sustainable consumption, in 12 countries of Asia.

The Asian Review on Sustainable Consumption was not officially published by the SC. Asia Project, but provided a base for the Asian Guidance Manual on Sustainable Consumption. This manual, published for public consumption by the Centre for Environment and Development in 2007 with the same findings, significantly differed from the approach of the European review, took a more holistic approach towards sustainable consumption and challenged the current approach and outlook of the United Nations 10 Year Framework of Programmes on Sustainable Consumption, popularly called the Marrakech process. It rejected the notion that eco-innovation or 'greening the consumer' should be the main focus of sustainable consumption in Asia, and essentially proposed that the primary goal of sustainable consumption should be to assure better quality of life for all. It suggested that sustainable consumption planning should enable wellbeing and ensure happiness for all people.

The follow-up in promoting national policies, action plans and implementation programs for sustainable consumption was not adequately monitored by the United Nations Marrakech process. As well, the interest of African and Latin American consumer and environmental NGOs in conducting similar regional reviews was not supported adequately. The process that created rapid awareness amongst all sectors in Asia including government, civil society, business and academia during 2004 to 2005 suddenly stopped without a proper follow-up plan for continuity, commitment and financial support from the UN Marrakech Secretariat. Therefore, a promising process to engage consumer activism in sustainable consumption went into a mode of sleep which to date has not been adequately awakened, despite some efforts to develop national guidelines for sustainable consumption based on the experience of the *Guidance Manual for Sustainable Consumption in Asia.*

References

AccountAbility & CI (2007) What assures consumers on climate change: Switching on citizen power. AccountAbility and Consumers International (CI), London

Anwar Fazal (2001) Consumers in the new millennium: back to basics. Federation of Malaysian Consumers Associations (FOMCA), Malaysia

Bentley M et al (2002) Tracking progress: implementing sustainable consumption policies. United Nations Environment Programme (UNEP), Division of Technology, New York

Consumers International (CI) (2001), Asia Pacific Office, Kuala Lumpur

Deborah J (2002) Consumer activism and corporate accountability: a history of the fair trade movement. J Res Consum, Sept 01

Diana MM (2004) From Cancun to Sao Paulo: the role of civil society in the international trading systems. Paper presented at the Afro-Asian Civil Society Seminar, Centre for International Trade, Economics and Environment (CUTS), New Delhi, 13–15 Apr 2004
Girimaji P (1999) Consumer rights for everyone. Penguin Books India, New Delhi
Mohd Hamdan Hj. Adnan (2000) Understanding Consumerism, Consumers Education, Malaysia
Hamilton C (2003) Growth fetish. Allen and Unwin, Australia
Kalam APJ, Rajan YS (1998) India 2020: a vision for the new millennium. Penguin Books India, New Delhi
Khor M (2000) Globalization and the South: some critical issues. Third World Network, Malaysia
Khor M (2002) Intellectual property, biodiversity and sustainable development. Third World Network/Zed Books, Malaysia/UK
Meadows DH, Meadows DL, Randers J (1992) Beyond the limits to growth: confronting global collapse, envisioning a sustainable future. Chelsea Green Publishing, UK
Rhodes S (2005) Social movements and free-market capitalism in Latin America: telecommunications privatization and the rise of consumer protest. SUNY Press, New York
Rutivi C (2008) Sustainable lifestyles/consumption from an African consumer perspective. Paper presented at the fifth African roundtable on sustainable consumption and production (ARSCP-5), Johannesburg, South Africa, 4–6 June 2008
Sahai S (2003) Genetically modified crops: a resource guide for the Asia Pacific. Gene Campaign, New Delhi
UN (2003) United Nations guidelines for consumer protection. UN Department of Economic and Social Affairs, New York
UNEP (2005) Advancing sustainable consumption in Asia: a guidance manual. United Nations Environment Programme (UNEP) /Earthprint, New York
Yunus M (2006) Nobel peace prize lecture. Explore 3(5):445–448, Oslo
Zoysa U de (2007) The Asian review on sustainable consumption. Report presented at the 3rd international expert meeting on sustainable consumption and production held in Stockholm, Sweden

Chapter 5
Sustainability in the Electricity Production and Consumption System – A Consumers' Perspective

Doris Fuchs and Sylvia Lorek

Introduction

The production and consumption systems (PACS) of electricity provide broad insights into possible actions and interventions in pursuit of sustainable development at different points in the life cycle from production and supply to consumption. More importantly for the context of the present book, electricity PACS allow particularly relevant insights regarding knowledge-to-action-gaps in sustainable development. Considerable knowledge exists on the sustainability challenges of electricity production and consumption. Not all of this knowledge is uncontroversial, of course, but in many respects we notice a sufficiently broad agreement among scientists on necessary and desirable changes. Knowledge on the various actors' and stakeholders' opportunities in improving the sustainability of electricity production and consumption is somewhat less precise. Still, even here, a substantial amount of information exists. Action, however, is missing, in many cases.

Knowledge-to-action gaps are particularly noticeable when it comes to consumers. Individuals can play a range of roles with respect to the composition of the electricity supply. Reisch and Micklitz's (2006) analysis of consumers' activities in the deregulated electricity market in Germany proposes a three-role model of consumers as market players, as citizens, and micro-producers in households and networks. In these roles, consumers take on different social and political identities; they are affected in different ways by (de)regulation of essential services and have different options for reacting to quality and price issues. In Chapter 3 in this volume, Reusswig concentrates on the power and influence of consumers working as citizen networks and organizing market power through investing in alternative (wind) energy production. In this chapter, however, we will concentrate on consumer choice in energy supply.

D. Fuchs
Professor of International Relations and Development at the University of Munster, Germany
e-mail: doris.fuchs@uni-muenster.de

S. Lorek
Sustainable Europe Research Institute, Head of Sustainable Consumption Research, Overath, Germany
e-mail: sylvia.lorek@seri.de

L. Lebel et al. (eds.), *Sustainable Production Consumption Systems: Knowledge, Engagement and Practice,* DOI 10.1007/978-90-481-3090-0_5,

Starting in the 1990s, many OECD countries liberalized their electricity markets. Germany, on which most of our analysis will focus, followed this trend in 1998. Third party access to the grid was introduced and consumers were provided with more choices regarding their energy supply. Besides increasing economic efficiency, this development was meant to allow consumers to exercise market power in support of renewable energy sources. Accordingly, green energy labels were created. In addition, NGOs campaigned for regulations requiring electricity suppliers to provide consumers with information on the composition of their electricity mix on the electricity bills. Despite these developments, consumer action in support of green electricity has been limited. This is the case, even though a general declared interest of German consumers in renewable energy sources, climate change and environmental issues as such, can be noticed. In other words, consumer "knowledge" has not been followed by action. Why not?

In this chapter, we aim to identify these consumer relevant knowledge-to-action gaps in German electricity PACs and their sources, considering broader developments in the European Union as a backdrop where relevant. In the course of this endeavor, we demonstrate that improved knowledge at the consumer level will only lead to marginal improvements in the sustainability of electricity PACS. We show that most of the legislation aimed at improving consumer information about the sources of their electricity supply is well intentioned but will not lead to a substantial transformation of the overall supply mix. This is partly the case due to the weaknesses in the legislation that facilitates disinformation as much as information. It is also the case, however, because this policy intervention is based on faulty assumptions about the sovereign and enlightened consumer and the appropriate distribution of responsibility for sustainability in electricity PACS. In short, the consumer does not receive the information in a form appropriate for inducing action. Further efforts targeted at information access and processing, as well as consumer interests and transaction costs will be necessary to induce consumers to effectively turn this knowledge into action. Specifically, politicians (as well as NGOs and scholars) need to realize limits to consumer sovereignty and activism that exist even in the case of well-informed consumers and intervene with the necessary stringency at more appropriate places in the system.

Our analysis is based on a review of the scientific and political literature on energy policy as well as sustainable consumption and production literature in general. In addition, we conducted in-depth interviews with experts from various stakeholder groups. From the data gathered, we draw information on political output, that is the current state of regulation on consumer information about the electricity supply in Germany and the European Union, and on the expected political outcome, that is changes in consumer choices regarding their electricity supply.[1]

[1] In the interest of space and given the complex dynamics at play we neglect the question of impact, that is the broader socio-economic consequences and implications of the policy here. For the analytical distinction between policy output, outcome, and impact see Easton 1965.

Given that part of the policy change we analyze is relatively recent, firm statistical results on its outcome are not available yet. Extensive information on consumer behavior in the sustainable consumption literature, however, strongly suggests that the outcome will be one of very limited changes in consumer choice in electricity supply even if the majority of consumers would support improvements in the sustainability of the supply mix in general. In our conceptual analysis of knowledge-to-action-gaps, we employ the AKIDA concept that delineates essential steps from attention and knowledge to interest, desire and action. We suggest that this concept is particularly suitable to highlight obstacles to action that would otherwise not become apparent.

The chapter is structured as follows. The next section will provide brief background information on the sustainability challenges involved in electricity production and consumption, thereby demonstrating the relevance of the field of inquiry for sustainable development. Section on Knowledge Availability, Creation, and Linkage in Electricity Production and Consumption Systems will turn to the questions of knowledge availability, creation, and linkage in electricity PACS. Given that much of the general knowledge on the sustainability challenges of electricity production and consumption is not all that controversial, this section will concentrate on the knowledge held by the various actors regarding their responsibilities and opportunities in influencing the sustainability of electricity PACS. Section on From Knowledge to Action, the core of the chapter, will examine the consumer related knowledge-to-action-gap in the sustainability of electricity production and consumption. First, the section will introduce the AKIDA concept and delineate its components and benefits. Secondly, it will apply the concept to German consumer information on their supply mix and delineate the relevant knowledge-to-action-gap. Section on Leverage then concludes the chapter by summarizing its results and implications and suggesting leverage points for sustainability in this field.

Background: Sustainability Challenges of Electricity Production and Consumption

Energy production and consumption present a core challenge for sustainability. Today, electricity is accounting for almost a quarter of the final energy demand worldwide. Future outlook shows electricity demand per capita to continue to rise steadily with the per capita GDP. It is expected to be 4 times higher in 2030 than in 2007 (International Energy Agency 2003).

Access to electricity and energy security are huge challenges for the social and economic dimensions of sustainability. This chapter's focus, however, will be on the environmental dimension. Nearly every kind of electricity production has negative impacts on the environment. They start with the mining of fossil fuels, which frequently is associated with major impacts on landscapes and the use of large amounts of energy to get the fuel out of the ground. Transporting the fuel again is related to high energy consumption and, in addition, carries the risk of ecologically disastrous accidents.

The most important and frequently discussed impacts are those of energy production in power plants. These effects are not limited to specific regions but cause problems on a global scale. The burning of fossil energy carriers – coal, mineral oil, gas, etc. – leads to emissions of CO_2, SO_2, and NOx. While the latter two can be reduced by technical solutions, CO_2 emissions from burning fossil fuels cannot be prevented and therefore play a crucial role in global warming.[2] Yet the attempt to reduce CO_2 emissions has reached specific importance, due to concerns about climate change. The use of nuclear energy carriers, in turn, creates radioactive waste with serious problems of safe storage.[3] In consequence, renewable energy carriers have become the focus of attention in pursuit of greater sustainability of electricity PACS, in particular the reduction of CO_2 emissions.[4]

A different perspective on the environmental implications of electricity PACS is provided by the sustainable consumption perspective. Consumers often tend to ignore the environmental impacts of their behaviour, because the use of electricity is invisible. The overall amount of electricity consumption, however, is the source of the environmental burden created. In consequence, reducing the overall amount of electricity consumed through improved efficiency as well as sufficiency measures has to be an environmental objective as well.

In sum, electricity PACS are associated with substantial environmental problems. Extensive knowledge on these environmental implications of electricity PACS is available. International and national official directives, guidelines and laws take it into account.[5] In those cases, where controversies exists, such as in the case of nuclear power, the lines between the different interests and perspectives are clearly drawn and a solution of the conflict is not likely to be a function of more knowledge.

[2]The possibilities of carbon dioxide capture and storage are still in the early phases of technological development and their potential for improving the sustainability of electricity generation from fossil fuels is highly controversial at this point.

[3]This is only the most serious risk associated with nuclear energy, of course.

[4]With the adoption of the *Kyoto Protocol*, the European Union committed itself to a reduction of greenhouse gas emissions of 8 percent of the 1990 level by 2012. Germany's internal targets are even more ambitious.

[5]For example: Directive 2001/77/EG; Directive 2003/54/EG; Green Paper KOM(2005) 265; Directive 2006/32/EG

In other words, the primary obstacles to improving the sustainability of electricity PACS are not a function of missing knowledge.

Knowledge Availability, Creation, and Linkage in Electricity Production and Consumption Systems

What knowledge is most needed for sustainability in the production-consumption system? Two types of knowledge come to mind here. First, we require an understanding of what the major challenges to sustainability arising from the electricity PACS are. As section on Background: Sustainability Challenges of Electricity Production and Consumption has pointed out, this knowledge is available to a substantial extent, and much of it is not that controversial, at least from an environmental perspective. In those cases, where controversies exists, such as in the case of nuclear power, the lines between the different interests and perspectives are clearly drawn and a solution of the conflict is not likely to be a function of more knowledge.

The second type of knowledge needed is that of actors regarding their own opportunities and responsibilities for improving the sustainability of electricity production and consumption. Given the range of actors involved – producers, suppliers, consumers, governments, NGOs, science, and media – there is a considerable spread in actors' opportunities to achieve the necessary improvement in the sustainability of electricity production and consumption. Knowledge on these actor-specific opportunities for intervention is available to a considerable extent as well. Yet, it is frequently not associated with the appropriate action.

Sustainability objectives in electricity production and consumption are traditionally associated with three key goals or guiding ideas: efficiency, sufficiency, and renewables. Efficiency approaches reduce the environmental burden through technical improvements. They reach from more efficient power generation via a change towards less CO_2 emitting fuels and combined heat power plants (CHP) to the use of appliances with higher efficiency grades. Often, those efficiency solutions are successful within their boundaries of reference only. Rebound effects generally render improvements in the efficiency of electricity consumption futile. As a consequence, sufficiency strategies, pursuing structural solutions for less energy consuming lifestyles, are of increasing importance. Finally, electricity from renewable energy sources offers the possibility of energy services with significantly reduced impacts on the environment.

For each of the three goals/guiding ideas, responsibilities for the range of stakeholders can be identified so that we arrive at a matrix of actors' responsibilities (see Matrix I; please note that the matrix contains examples for illustrative purposes only and does not aim for comprehensiveness).

Matrix I: Actors' Responsibilities

	Decision makers within the production–consumption system			Knowledge providers[6]		
Actors goals	Producers	Suppliers	Consumers	NGOs	Science	Media
Efficiency	Increase efficiency of power plants	Demand Energy form BHKW Support consumer investments in energy efficient appliances	Demand electricity with low CO2 content Demand appliances with higher efficiency grades	Provide information and lobby for efficient energy production, supply, and appliance standards	Develop technical solutions, raise awareness	Provide information about efficient energy production and supply
Sufficiency		Prohibit tariff structures that support high energy consumption	Reduce electricity demand through behavioral changes	Provide information on low energy consuming lifestyles	Develop social solutions, raise awareness	Support vision building on low energy consuming lifestyles
Renewables	Produce electricity from renewable sources: wind, solarv	Offer special eco-tariffs, change portfolio of the company	Ask for/change to electricity from renewable sources	Provide information and lobby for regulations supporting renewables and demand for electricity from renewable sources	Develop technical solutions, raise awareness	Provide information about renewable energy and possibilities of change

[6]The lines between the "decision makers within the production and consumption system" and the "knowledge providers" frequently are blurred, of course.

Matrix II: Regulators' Responsibilities

Actors goals	Decision makers within the production–consumption system			Knowledge providers		
	Producers	Suppliers	Consumers	NGOs	Science	Media
Efficiency	Define legislative and regulative requirements Tax production depending on CO_2 emissions	Tax depending on CO_2 emission of the products	Define standards for appliances Define requirements for information disclosure	Fund and support campaigns Integrate NGOs in consulting and decision making bodies	Improve funding for sustainability research Implement results from sustainability research	Strengthen knowledge provision for sustainability in public media Regulate misleading advertising
Sufficiency		Regulatory / requirements for tariff structure	Individual CO_2 certificates			
Renewables	Define minimum requirements for energy mix Provide subsidies for electricity production from renewable sources	Liberalize grid access Define requirements for eco-tariffs	Define requirements for disclosure of energy sources, standardize labels, provide subsidies/tax relief.			

Regulators are a special actor, in this context, as they can intervene with respect to the behavior of all actors in pursuit of all goals/guiding ideas.[7] Thus, they have not been included in the above matrix. Ideally, this matrix should be three-dimensional then, providing space for regulators' options regarding each cell in the third dimension. As a simplified version, one can also conceive a similar matrix that delineates regulators' options for the various cells (Matrix II).

Clearly, we cannot discuss all matrix cells even for just one of the matrices in the limited context of this paper. The consumer column is the most relevant one, and therefore we will focus on that from now on. Here, we would argue that the fundamental knowledge about the unique importance of renewable energy sources already exists. Moreover, as several recent surveys have pointed out, close to 90% of citizens in Germany have displayed their preference for energy from renewable sources (Forsa 2006). The main arguments provided by consumers for these preferences are sustainability considerations (44%) as well as those of long term energy security (25%) (Europressedienst 2005).

What do consumers know about their own opportunities for supporting renewables? In 1998, German electricity markets were liberalized. Since then, consumers in Germany have been able to choose between different electricity suppliers offering various supply mixes. Awareness of this option is relatively high as intensive advertising campaigns were launched by providers of cheap electricity (mostly produced in nuclear power plants).

Yet consumer action has been limited. Less than 7% of German consumers have changed their supplier; and most of them turned towards cheaper providers (Bundesnetzagentur 2008). Only a small share of the consumers having changed suppliers demonstrated an interest in less CO_2 emitting and nuclear free electricity by turning to providers of "green" electricity.

To some extent, this lack in consumer action could be explained by a lack of knowledge on the part of consumers. First, the four competing "green electricity" labels were launched, which led to consumer irritation where orientation was needed. Secondly, an important part of the knowledge required for consumers to make appropriate choices was missing completely: reliable information on one's energy mix. Consumers could purchase electricity contracts just for renewable power or with a certain percentage of renewables. Unless they did so and explicitly sought information on their supply mix, however, they had little information about the renewables as well as coal, gas, or nuclear power shares in the energy mix of their suppliers.

This knowledge gap was supposed to be closed by new legislation requiring electricity suppliers to inform their customers about the energy mix in their supply. In 2005, the German Energy Industry Act adopted a relevant EU Directive[8] from 2003.

[7]Of course, regulators themselves also are susceptible to the influence of social forces, for example from business and civil society. Chapter 3 in this book written by Fritz Reusswig analyzes this relationship and its importance in pushing electricity production and consumption systems towards sustainability.

[8]See: EnWG § 42 and EU Directive 2003/54/EG

The Act specified electricity providers' obligation to provide information about the share of energy carriers in their overall energy mix and the related implications for the environment, specifically the CO_2 emissions and nuclear waste. For better comparison, data on the German average have to be added. With the latter condition, the German legislation reaches beyond the requirements of the EU Directive.

Still the legislation on disclosure requirements suffers from major weaknesses as NGOs point out (Greenpeace 2006). The presentation of information on the bills frequently tends to be not intentioned to raise consumers' interest and desire for action. Information, for example, can be given on the back side of the bill. Likewise, this information can be provided as pure text, a method that often is associated with the least informational content, as a marketing study conducted in the field of regulations pointed out (Utilitias 2004). Thus, one may suspect that little consumer action is to be expected without sufficient improvements in the requirements for disclosure of the energy supply mix. Example 1 illustrates an insufficient adaptation of the directive while example 2 was recommended from the EU project "Consumer Choice and Carbon Consciousness for Electricity (4C Electricity)"

Example 1: Insufficient Adaptation[9]

The information is provided in hidden and unattractive ways, for example in a footnote on the backside of the bill as shown in the case of Aggertal GmbH.

Example 2: Recommended Adaptation

As we suggest below, however, little consumer action is to be expected even if the disclosure requirements are improved. Specifically, we argue that improved knowledge in combination with the opportunities provided by liberalization will not foster substantial consumer action in support of renewables in a major way due to the existence of a substantial knowledge-to-action gap on the part of consumers.

[9]Die von der Stromversorgung Aggertal GmbH im Jahr 2005 gelieferte elektrische Energie setzt sich aus folgenden Energieträgern zusammen (Durchschnittswerte Deutschland zum Vergleich – Quelle VDEW):

Eighteen percent (29%) Kernkraft, 65% (60%) fossile und sonstige Energieträger (z.B. Steinkohle, Braumkohle, Erdgas) und 17 % (11%) Erneuerbare Energien. Damit sind folgende Umweltauswirkungen verbunden: 0.0005 g/kWh (0.0008 g/kWh) radioaktiver Abfall sowie 413 g/kWh (514 g/kWh) CO2-emission.

[Electricity provided by Aggertal GmbH in 2005 was from the following sources (German average): 18% (29%) nuclear energy, 65% (60%) fossil sources (e.g. coal, brown coal, natural gas), and 17% (11%) renewable energy. This caused the following environmental burden: 0.0005 g/kWh (0.0008 g/kWh) nuclear waste and 413 g/kWh (514 g/kWh) CO2 emissions].

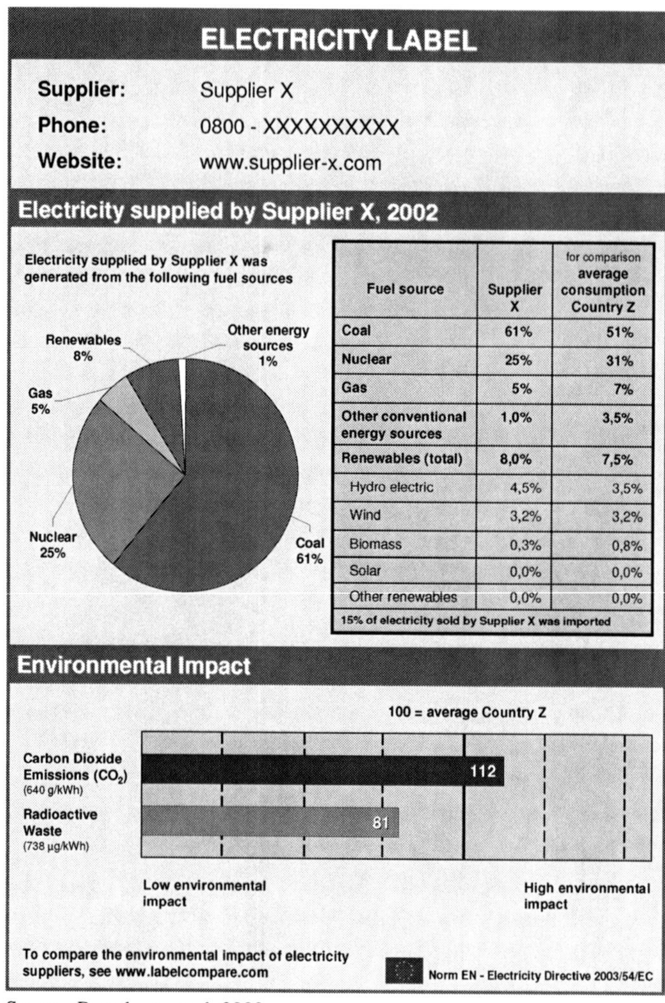

ELECTRICITY LABEL

Supplier: Supplier X
Phone: 0800 - XXXXXXXXXX
Website: www.supplier-x.com

Electricity supplied by Supplier X, 2002

Electricity supplied by Supplier X was generated from the following fuel sources

Renewables 8%
Other energy sources 1%
Gas 5%
Nuclear 25%
Coal 61%

Fuel source	Supplier X	for comparison average consumption Country Z
Coal	61%	51%
Nuclear	25%	31%
Gas	5%	7%
Other conventional energy sources	1,0%	3,5%
Renewables (total)	8,0%	7,5%
Hydro electric	4,5%	3,5%
Wind	3,2%	3,2%
Biomass	0,3%	0,8%
Solar	0,0%	0,0%
Other renewables	0,0%	0,0%
15% of electricity sold by Supplier X was imported		

Environmental Impact

100 = average Country Z

Carbon Dioxide Emissions (CO_2) (640 g/kWh) 112
Radioactive Waste (738 µg/kWh) 81

Low environmental impact
High environmental impact

To compare the environmental impact of electricity suppliers, see www.labelcompare.com

Norm EN - Electricity Directive 2003/54/EC

Source: Boardman et al. 2003

From Knowledge to Action

The AKIDA Concept

To explore the knowledge-to-action-gap relevant for our case, we employ the AKIDA concept, which delineates interest and desire as important steps between knowledge and action. This concept builds on the AIDA (attention, interest, desire, action) concept traditionally used in marketing, which depicts the steps advertisers need to get consumers to carry out when selling a product.[10] In applying this concept to electricity consumption, we fill it with information provided by an economic perspective, which highlights the relevance of transaction costs and collective action.

In the marketing concept, the question of knowledge is not addressed, as "knowledge" of and about the product tends to be packaged with emotional enticements as part of the message that is being transmitted. For our purposes, however, knowledge is an important element. In the political process, a lack of knowledge frequently is assumed to be the cause of the problem; and if the knowledge exists, observers wonder about a lack of appropriate action. Thus, knowledge needs to be considered an independent step in the process.

In order for knowledge to be gathered, attention must be paid to a particular question as awareness of a problem arises. Thus, knowledge is not the first step in this chain. In our case, however, knowledge exists and thus we will not address the question of who and what can create attention for a problem here. Instead, we will focus on the part of the chain between knowledge and action.

The AKIDA concept suggests that interest and desire are the steps that need to be completed successfully, if action is to result. Interest here is the more general attitude that provides the ground for the specific individual preparedness and desire to take action. How far can these steps help us understand a potential knowledge-to-action-gap for consumer support of electricity from more sustainable sources? First, these steps remind us that knowledge is not a sufficient condition for action. Secondly, they highlight that a general interest in a "product" is necessary, that is a general support for renewables in our case, as well as a specific and pointed desire to acquire the product and benefit from its ownership and use. Can we expect these intermediate steps to be taken when it comes to consumers and the electricity supply mix?

The Knowledge-to-Action-Gap in Political Consumerism in Electricity PACS

Unfortunately, much of what we know about consumption and political consumerism today suggests that the knowledge-to-action-gap will not be closed in the case of demand for electricity from more sustainable sources, even if consumers have perfect information on their supply mix. Admittedly, scholars and practitioners often proclaim a new awareness and interest in the environmental and social effects of consumption by consumers:

[10]The use of the AIDA concept is controversial in marketing today, due to its simplification of the psychological and social determinants of consumption. It is useful for our purposes, however, in that it draws our attention to the intermediate steps necessary before action is to be expected.

It is becoming more and more evident that consumers are increasingly interested in the "world that lies behind" the product they buy. Apart from price and quality, they want to know how and where and by whom the product has been produced. This increasing awareness about environmental and social issues is a sign of hope. Governments and industry must build on that (Klaus Töpfer, 23 August 1999, UNEP News Release NR 99-90).

Likewise, surveys show a high ratio of consumers concerned about the impacts of their consumption behaviour (Bentley and de Leeuw 2000). According to a global consumer survey by UNEP, 93% of consumers are, on average, aware of the impact of their consumption patterns on the environment, 60% are quite concerned, and more than 30% always or mostly consider the processes that lie behind the product (Bentley 2000). Some consumers do adopt alternative lifestyles based on ideas and values such as simple living, the diffusion of which is fostered by global communication networks (Georg 1999; Maniates 2002; Schor 1998).

A political focus on opportunities for consumer activism thus appears to suggest itself. Such opportunities were meant to be promoted with the liberalization of the electricity market (besides economic objectives such as competition and price reduction, of course) and the disclosure requirements discussed above. Likewise, the Commission's Green Paper on Energy Efficiency "Doing more with less" has demanded that "shortcomings in information and training of consumers and the general public" have to be overcome (European Commission 2005).

However, environmental, social, or sustainability values are competing with a multitude of other criteria in their influence on consumption decisions (Fuchs and Lorek 2001). After all, consumers make their consumption decisions in certain given social, economic, and technological contexts at micro- and macrolevel (Michaelis and Lorek 2004; Reisch and Scherhorn 1999; Røpke 1999, 2001). Thus, while some commentators postulate the existence of consumer sovereignty in the global market place – a view implying that it is solely consumers who determine production patterns – few scholars working on sustainable consumption share this view to its full extent. In fact, it is well known that surveys frequently indicate higher levels of perceived "green" behaviour by individuals than their actions reflect.[11]

In fact, when it comes to action, there is ample evidence that sustainability criteria often rank low compared to competing aims. After all, consumption objectives include the wish to express status, define one's identity, or establish belonging, for instance. The tendency to satisfy such needs on the basis of material consumption increases in a globalized world, in which traditional social networks are increasingly disappearing, interpersonal contacts become briefer and more superficial, and personal insecurity is rising (Scholte 2000). Moreover, given increasing demands on consumers' attention, time, and budget, it has to be

[11]Likewise, empirical research has shown that consumers faced with a task of reducing their energy consumption, for instance, are willing to make small sacrifices only and generally fail to achieve the required level (Gatersleben and Vlek, 1998).

acknowledged that factors such as convenience and price matter more and more in decisions about consumption.[12]

Furthermore, in the flow of global communications, "sustainability" messages are overpowered by opposing ones. The advertising industry estimates that every consumer has thousands of brand contacts per day. Even those watching the development and diffusion of alternative lifestyles with hope will have to acknowledge that these are still marginal and likely to remain so relative to the overall focus on the societal trends supporting unsustainable consumption (Georg 1999).[13]

Still, with respect to renewables, we can identify the attitude of general interest, which the AKIDA concept identifies as necessary for the overcoming of the knowledge-to-action-gap among consumers. As pointed out above, surveys report that close to 90% of German consumers identify renewables as their favorite energy source (Forsa 2006). In the European Union, 46% of those surveyed report a willingness to pay 10% more for a more sustainable energy mix in the electricity supply (European Commission 2006a). Forty-nine percent of women and 48% of men consider solar energy the most important contributor to the future security of the energy supply (ibid.).

When it comes to action, however, the next important step in the AKIDA concept, specific individual desire to act on the basis of the information about one's energy mix is less likely to exist. Environmental consumer research has shown that environmentally superior consumption choices tend to require the perception of a direct personal benefit. In the case of the energy mix of one's electricity supply, however, such a benefit is likely to exist only for a small group of consumers, usually identified as the "eco-niche," "deep green" consumers or "ecological innovators".[14] The benefits of switching to green electricity for this group are likely to be relatively high and outweigh the perceived costs, as members of the group consider environmental action necessary, environmental consciousness is part of their identity, and/or environmental action also tends to satisfy their need for participation.

For the majority of consumers, the perceived direct personal benefit from switching to a more sustainable energy mix is likely to be too low to lead to action. This is especially the case, since the socio-economic and technological contexts of the relevant consumption decisions foster inertia and are associated with considerable

[12]See Fuchs and Lorek 2005 for a discussion on the even larger obstacles to sustainable consumption in the context of sufficiency issues.

[13]The situation is different in the case of food scares and associated health concerns, which have caused (frequently temporary) changes in food consumption patterns in some countries.

[14]Empirical studies on sustainable consumption have identified a range of consumption styles, which are characterized by different consumer preferences and abilities resulting from the combination of subjective and societal determinants of consumption (Prose 2000; Villiger et al 2000). Schultz (2000a, b), for example, distinguishes between four target groups for policy intervention with respect to the sustainability characteristics of household consumption choices: an environmentally oriented consumption style, the status oriented elite, a consumption style that is not responsive to environmental or social messages, and a large remaining group of consumers, who will include environmental criteria in their consumption choices under certain conditions.

transaction costs. In the case of electricity, path dependency means that consumers are locked into the socio-economic context and practice associated with the previous technology regime (Fuchs and Arentsen 2002). After all, consumers are used to receiving their electricity from a monopoly provider with no room and need for making choices (Truffer et al. 2001). This has several consequences. First, consumers are used to regarding decisions about the environmental impact of electricity generation and its future trajectory as the responsibility of the regulator and the utilities. An EU survey even found that 47% of those interviewed saw the primary responsibility for the sustainability characteristics of the electricity supply at the European level of decision-making (European Commission 2006b). Secondly, consumers have no experience with purchasing electricity in a market environment. Where does one purchase electricity and from whom? Where does one get the necessary information, and what is the important information? Given such "ignorance" and uncertainty, many consumers find consumer choice overwhelming (Wiser et al. 1999), and choose to stay with their previous monopoly suppliers who are still benefiting from the aura and history of secure electricity supply.[15] Besides a lack of factual knowledge about the supply side, consumers are starting with a lack of knowledge about their own needs and preferences. Thus, consumers are frequently rather uninformed about their own electricity consumption, not just in terms of its environmental impact but also in terms of cost differentials. This is quite understandable since such information was of little relevance as long as little choice and room for change existed. Finally, consumers willing to switch to green electricity face collective action problems. After all, the individual choice has little impact on the overall environmental impact of electricity generation. In sum, high transaction costs and collective action problems are likely to prevent even stringent disclosure requirements for annual electricity bills from leading to substantial consumer action.

In the case of the liberalized electricity market in Germany, efforts to reduce transaction costs have partly failed and are further characterized by asymmetry in the provision of information to consumers. As pointed out above, they have failed in so far as four environmental labels each with an additional set of grades compete in the market. Some of these labels have been created by environmental NGOs while some at the behest of the large electricity producers and suppliers. Thus, choosing an improved energy mix is actually quite a demanding task. Transaction costs, furthermore, are characterized by an asymmetry that favors the large electricity suppliers (with large shares of coal and nuclear power plants) which have the biggest marketing budgets as well as the most extensive retail networks.

[15]Reducing transaction costs for consumers means also the reduction of uncertainty and perceived risk. Thus, consumers need to be assured of consumer protection, especially with respect to contract reliability and consequences of supplier failure. Likewise, consumers need to know to what extent they can trust information on producer "performance characteristics "such as the environmental impact of the generation of that electricity. Note, however, the complex relationship between credibility and accessibility of information.

In their case, the message of "low price" is frequently communicated in combination with "easy to get".[16]

There is, of course, always the example of the Brent Spar to counter such a pessimistic evaluation of the potential for political consumerism. Yet, the differences in the situations should not be underestimated. The Brent Spar effect was an immediate and short term reaction to the scandalization of a traditional practice in the oil industry that clearly had the potential to grab more attention than the publication of one's energy supply mix on a yearly bill. The opportunities for reaction were rather simple: go to another gas station (generally next door or less than a couple of kilometres away). It was a frequent action that did not require the gathering of new information on processes and suppliers, and that could be taken every day and did not require the cancellation of (potentially long term) contracts. Finally, in its brevity and intensity as a concerted action it provided consumers with the benefit of "participation" in a political course that long-term decisions made from the individual home have difficulty to match.[17]

In sum, it is highly unlikely that even improved provision of knowledge about one's relative CO_2 emissions and the make-up of one's electricity supply will lead to major improvements in consumption decisions regarding electricity. Major sources of a gap between knowledge and action can be identified: the lack of the perception of a personal benefit for a large share of the population, the high transaction costs for switching to more sustainable energy supply mixes, the asymmetry in information availability and transaction costs, the collective action problems associated with political consumerism, and the perception that ensuring a sustainable energy supply mix is the task of government. Some of these obstacles to action may be reduced. Thus, improving the accessibility and perceptions of the reliability of information and associated measures to reduce transaction costs may help a bit. More communication targeting emotional attachments and personal benefit could potentially reach some consumers.[18] Likewise, improved knowledge on the urgency of the issue may cause a few more consumers to switch to more sustainable electricity supply mixes. For the most part, however, such measures are likely to fail in fostering a substantial improvement in the sustainability of the electricity supply, due to the major obstacles to sovereign and political consumerism pointed out above.

[16]In other cases, they also transmit "environmental" messages quite successfully, selling "water power," which is produced by old, large scale water power projects and does not improve the current energy supply mix, at a premium.

[17]There have, however, been efforts to achieve such experiences of participation in the form of concerted actions fostered by a German TV soap, for instance.

[18]The "traditionals" in Schultz' (2000a, b) studies, for instance, tended to respond to messages of health, regional context, and specific rather than general environmental benefit (Fuchs and Arentsen 2002). Likewise, "the privileged," as Schultz calls them, can potentially be convinced to switch to green electricity, if they perceive the latter as valuable in providing status due to its high tech quality and expensiveness (e.g. solar power) (ibid.).

Leverage

In this chapter, we have aimed to demonstrate that current problems regarding the sustainability of the electricity system do not result from a lack of knowledge for the most part – at least not lack of information on the consumer side. Consumers could be provided with more appealing and 'interesting' information in the sense of the AKAIDA concept regarding the environmental implications of their supply mix. Such improved knowledge would require government intervention in terms of prescribing the more frequent, accessible and poignant provision of the relevant information by electricity suppliers. In addition, governments and/or NGOs could try to foster the willingness of consumers to take on responsibility regarding the sustainability impacts of their electricity consumption by reducing transaction costs and attempting to better communicate a "personal benefit."

Even with such improved information, this particular knowledge-to-action gap would remain, however. The obstacles to political consumerism created by the general contextual constraints on consumption decisions, collective action problems, and transaction costs would remain too high, even if the latter were reduced.

Politicians and bureaucrats appear to be aware that efforts to improve the sustainability of electricity PACS cannot rely on consumer decisions alone. We suggest that their expectations of the likely sustainability contribution by consumers as well as the effectiveness of more information provision should be even lower than they are. In fact, the main knowledge-to-action-gap appears to exist between the awareness of consumers' limited influence and the lack of sufficient governmental intervention on the supply side.

Many European governments do pursue a mixed strategy when it comes to the energy mix in electricity PACS, targeting both the demand and the supply side. The German government, in particular, has received substantial acclaim for its support of renewables through subsidies and access guarantees. Even here, however, government intervention on the supply side is insufficient. Such interventions not only need to include regulations regarding (increasing) minimum shares of renewables in the energy mix of electricity production, sufficiently high subsidies, as well as access provisions but also such interventions need to be targeted stringently, as if they were the only interventions taken.

References

Bentley, M (2000) Global consumer class report: in search of common ground for common good. United Nations Environmental Programme (UNEP), Paris

Bentley M, de Leeuw B (2000) Sustainable consumption indicators. Report for UNEP DTIE. United Nations Environmental Programme (UNEP), Paris

Boardman B et al (2003) Consumer choice and carbon consciousness for electricity (4CE). Final report. Environmental Change Institute, UK

Bundesnetzagentur (2008) Monitoringbericht 2008 [Monitoring report 2008], Bonn

Easton D (1965) A systems analysis of political life. Wiley, New York

European Commission (2001) Directive 2001/77/EC of the European Parliament and of the Council on the promotion of electricity produced from renewable energy sources in the internal electricity market, Brussels
European Commission (2003) Directive 2003/54/EC of the European Parliament and of the Council concerning common rules for the internal market in electricity, Brussels
European Commission (2005) Green Paper on energy efficiency or doing more with less. COM(2005) 265 final, Brussels
European Commission (2006a) Directive 2006/32/EC of the European Parliament and of the Council of 5 April 2006 on energy end-use efficiency and energy services, Brussels
European Commission (2006b) Attitudes about energy. Special Eurobarometer 247, Brussels
Europressedienst (2005) Akzeptanz Erneuerbarer Energien in Deutschland im Rahmen der Energiepolitik der Zukunft [acceptance of renewable energies in Germany in the context of an energy policy for the future], Bonn
Forsa (2006) Einschätzung der Sicherheit der Energieversorgung in Deutschland [estimations on energy security in Germany], Berlin
Fuchs D, Arentsen M (2002) Green electricity in the market place: the policy challenge. Energy Policy 30(6):525–538
Fuchs D, Lorek S (2005) Sustainable consumption governance – a history of promises and failures. J Consum Policy 28(3):261–288
Gatersleben B, Vlek C (1998). Household consumption, quality of life, and environmental impacts: a psychological perspective and empirical study. In: Noorman KJ, Schoot Uiterkam T (eds) Green households? Domestic consumers, environment, and sustainability. Earthscan, London, pp 141–183
Georg S (1999) The social shaping of household consumption. Ecol Econ 28:455–466
German Ministry of Justice (2005) Energiewirtschaftsgesetz (EnWG) [Energy Industry Act], Berlin
Greenpeace (2006) Stromkennzeichnung – Tranzparenz oder Verbrauchertäuschung [Electricity disclosure – transparency or deception of consumer]. Greenpeace, Hamburg
International Energy Agency (2005) World Energy Technology Outlook to 2050. OECD/IEA, Paris
Lorek S (2001) Household energy and water consumption and waste generation – German Case Study. Overath, OECD, p 134
Lorek S, Lucas R (2003) Towards sustainable market strategies – a case study on eco-textiles and green power. Wuppertal Institute, Wuppertal
Maniates M (2002) In search of consumptive resistance: the voluntary simplicity movement. In: Princen T, Maniates M, Conca K (eds) Confronting consumption. MIT Press, Cambridge
Michaelis L, Lorek S (2004) Consumption and the environment in Europe: trends and futures. Danish Environmental Protection Agency, Copenhagen
Prose, F (2000) Auswirkungen von Konsumstilen im Energiebereich. Paper presented at the BAUM workshop, Berlin, 15–16 February 1999
Reisch L, Scherhorn G (1999) Sustainable consumption In: Dahiya SB (ed) The current state of economic science, vol 2. Spellbound, Rhotak, pp 657–690
Reisch L, Micklitz HW (2006) Consumers and deregulation of the electricity market in Germany. J Consum Policy 29(4):399–415
Røpke I (1999) The dynamics of willingness to consume. Ecol Econ 28:399–420
Røpke I (2001) New technology in everyday life – social processes and environmental impact. Ecol Econ 38:403–322
Scholte JA (2000) Globalization: a critical introduction. Macmillan, Basingstoke
Schor J (1998) The overspent American; upscaling, downshifting, and the new consumer. Basic Books, New York
Schultz I (2000a) Nachhaltige Konsummuster: Bilanz und Perspektiven aus Sicht der Sozialwissenschaften. Paper presented at the BAUM workshop Nachhaltige Konsummuster. Möglichkeiten der Umweltkommunikation, Berlin, 15–16 February 2000

Schultz I (2000b) Entwicklung von zielgruppen- und konsumstilspezifischen Ökologisierungstrategien/ neuen 'Öko'-Angeboten. Paper presented at the BAUM workshop Nachhaltige Konsummuster Möglichkeiten der Umweltkommunikation, Berlin, 15–16, February 2000

Truffer B, Markard J, Wustenhagen R (2001) Eco-labeling of electricity strategies and tradeoffs in the definition of environmental standards. Energy Policy 29(11):833–945

United Nations Environment Programme (UNEP) (1999) UNEP News Release NR 99-90. Klaus Töpfer, 23 August 1999, Nairobi

Utilitias (2004) Stromkennzeichung–Überprüfung verschiedener Gestaltungsvorschäge. Utilitas Forschung, Darmstadt

Villiger A, Wüstenhagen R, Meyer A (2000) Jenseits der Öko-Nische [Beyond the eco-niche]. Birkhäuser, Basel

Wiser R, Fang J, Porter K, Houston A (1999) Green power marketing in retail competition: an early assessment. Working Paper LBNL-42286. Ernest Orlando Lawrence Berkeley National Laboratory, Berkeley, CA

Chapter 6
Agrofuels in Thailand: Policies, Practices and Prospects

Rajesh Daniel, Louis Lebel, and Shabbir H. Gheewala

Introduction

Agrofuels, defined here as agriculture-based liquid transportation fuels, have been widely promoted as a renewable, sustainable, energy source which could reduce dependency on fossil-fuel imports. But the evidence of current and potential future contributions to sustainability is modest and mixed, varying with crops, places, markets and societies.

In Thailand, agrofuels comprise ethanol or gasohol and biodiesel. Molasses, a by-product of sugar manufacture, is fermented to produce ethanol.[1] One ton of sugarcane yields about 45 kg of molasses out of which 10 L of ethanol can be produced (Gonsalves 2006). About 90% of ethanol in Thailand comes from molasses and the remaining from cassava (APEC 2008). Ethanol blended in gasoline (petrol) is called gasohol. If it contains 10% ethanol, it is called E10 (Gonsalves 2006). In 2007, Thailand's ethanol production was 193 million litres.

Biodiesel is manufactured by the "transesterification" of vegetable oil and then blended with diesel. B2 is 2% biodiesel with 98% petroleum diesel; a 10% blend with diesel is called B10 (Gonsalves 2006). Palm oil (from oil palm plantations mostly in the southern region) is the main raw material for biodiesel in Thailand.

R. Daniel (✉) and L. Lebel (✉)
Unit for Social and Environmental Research, Chiang Mai University, Chiang Mai, Thailand
e-mail: rajesh@sea-user.org
e-mail: louis@sea-user.org

S.H. Gheewala
Joint Graduate School of Energy and Environment (JGSEE), King Mongkut's University of Technology, Bangkok, Thailand
e-mail: shabbir_g@jgsee.kmutt.ac.th

[1] Sugarcane is grown mainly in the central and northern regions of Thailand in two main planting seasons: from February through to April if it is to be irrigated, and from April to May if it is rain-fed cultivation. In the northern region, planting season runs from October to November and is mostly rain-fed cultivation.

L. Lebel et al. (eds.), *Sustainable Production Consumption Systems: Knowledge, Engagement and Practice*, DOI 10.1007/978-90-481-3090-0_6,

One hectare (ha) of palm oil can produce 4–5 t of crude palm oil which is five to ten times more than the yield of any commercially grown oil crop (Gonsalves 2006). In 2007, Thailand's biodiesel production was 58 million litres.

In this chapter we assess agrofuel policies in Thailand focussing on the period 1990–2008. The chapter is organized as follows: first we review the key policies promoting agrofuel use in Thailand noting important similarities and differences with initiatives elsewhere in the world; then three related sections analyze the evidence about environmental, economic and social issues; we then end with synthesis drawing conclusions about priority policy needs for Thailand.

Agrofuel Policies

Policies and Programmes

The Ministry of Energy (MoE) of the Thai government has established a number of policies and set many long-term targets to maximize the use of agrofuels in the transportation sector. For now, it is too early to tell how well these policies have been working. Some targets – such as the planned oil palm expansion – already appear too ambitious.

The MoE targeted increasing renewable energy utilization from the level of 0.5% in 2003 to at least 6% of total energy demand by the year 2011 (Sarochawikasit 2006). Biomass-based energy is targeted to provide more than 60% share (Fungtammasan 2008; Nunt-Jaruwong 2008).

The Thai Ministry of Energy and its Department of Alternative Energy Development and Efficiency (DEDE) have been the main proponents of measures introduced for an "enabling environment for agrofuels development" (MoE 2007). A central agency called the "National Biofuels Committee" is responsible for developing and implementing a "Biofuels Strategic Plan" developed by the Ministry of Energy with support from key government ministries such as finance, agriculture, industry. The financing and project development mechanisms set up through Special Purpose Vehicles (SPV)[2] and financial institutions for project implementation are also equally important (Gonsalves 2006).

The main agrofuels policy plank of the Thai government is a two-phase gasohol programme (Table 6.1). Phase 1 (2004–2006) targeted three ethanol plants to come on-stream by the end of 2006 and increase the production capacity to 1.1 million litres/day. In this phase, the government mandated that all of its vehicles use gasohol and formulated policies on the use of gasohol in high-performance vehicles.

[2] SPV is a legal entity (usually a limited company of some type or, sometimes, a limited partnership) created to fulfill narrow, specific or temporary objectives. SPVs are typically used by companies to isolate the firm from financial risk. A company will transfer assets to the SPE for management or use the SPE to finance a large project thereby achieving a narrow set of goals without putting the entire firm at risk (http://en.wikipedia.org/wiki/Special_purpose_vehicle).

Table 6.1 Selected agrofuels policies and programs of the Thai government

Policy or program	Period	Objectives
Gasohol programme phase I	2004–2006	Three ethanol plants by end of 2006
		All government vehicles use gasohol
Gasohol programme phase II	2007–2012	Licenses to 18 new biodiesel plants to increase total installed capacity By 2012, all petrol consumed will be gasohol 95 by law
"Strategic Plan on Biodiesel Promotion and Development"	January 2005 onwards	Replace 10% of diesel consumption by increasing palm oil cultivation, and promoting community-based and commercial biodiesel production in 2012
		Introduced a B2 mandate in February 2008
Incentives	Since 2003	US$50 million for palm oil cultivation in the southern provinces
	2008	Subsidy of 0.30 baht/L from the State Oil Fund for B2 manufacturers
		Three billion baht in soft loans to oil palm growers

A total of 45 ethanol producers have registered with authorities for a combined production capacity of 12 million litres/day but as of 2007 only 11 factories are in production with daily output of 1.57 million litres. Output by 2009 was projected to rise by another 2.5 million litres/day as eight new plants start production.[3]

In Phase 2 (2007–2012), the government awarded licenses to 18 new biodiesel plants to further increase total installed capacity to 3million litres/day by 2012. Of the 18 new plants in Phase 2, 14 will use molasses feedstock and the remaining four will be cassava-based. By 2012, all petrol consumed will be gasohol 95 by law (Gonsalves 2006).

For developing biodiesel, the Thai government announced the "Strategic Plan on Biodiesel Promotion and Development" in January 2005 (Table 6.1). The plan aims to replace 10% of diesel consumption by increasing palm oil cultivation, and promoting community-based and commercial biodiesel production in 2012. Moreover, the government introduced a B2 mandate in February 2008 to require the production of approximately 420,000 t of biodiesel/year (APEC 2008).

B2 manufacturers can avail themselves of subsidy of 0.30 baht/L from the State Oil Fund. The government is also making available 3 billion baht in soft loans to farmers growing oil palm. The demand for crude palm oil and stearin (palm oil by-product) is estimated to be 492 million litres/year by 2010 as compared to only 31 million litres in 2006 (Preechajarn et al. 2008). Also, the government has budgeted US$50 million for palm oil cultivation in the southern provinces, an area also designated as a special Board of Investment (BOI) zone (Griggers 2004). The government is also supporting research into other crops like jatropha (APEC 2008). The government has agreed to approve US$32.5 million budget for research and development of biodiesel projects during 2005–2008. $20 million baht will be provided

[3] http://www.thebioenergysite.com/news/1299/thailand-an-ethanol-centre.

for Jatropha seedling and other feedstock, and $12.5 million for research, development and administration of biodiesel applications (IJO 2007).

The Department of Alternative Energy Development has opened a community biodiesel plant in Chiang Mai that uses waste oil for producing biodiesel that can be used in light duty (LD) diesel vehicles (Pleanjai et al. 2009b). The pioneering biodiesel project was done under the Royal Chitralad Projects of His Majesty the King of Thailand (RCP 2006). Other projects that followed were: bio-diesel produced from refined palm oil of the Department of the Royal Naval Dockyard; bio-diesel produced from waste (stearin) of palm oil industries and crude palm oil under the research projects of Agricultural Research Development Agency (ARDA 2005).

Diesel consumption is currently at 50 million litres/day and is expected to rise to 85 million litres/day by 2012. Thailand plans to have installed biodiesel production capacity of 8.5 million litres/day by 2012 so that B10 (10% biodiesel in diesel) can meet the total national diesel demand. By 2012, the ethanol and biodiesel programs are expected to save $325 and $675 million, respectively, each year for the national economy (Gonsalves 2006). At present, Thailand has nine biodiesel plants with a total production capacity of 655 million litres annually (APEC 2008).

Tax Breaks and Subsidies

Economic measures are in place to increase investments in the production and use of ethanol. These include a Board of Investment (BOI) privilege for a fuel ethanol plant, a waiver on the excise tax for the ethanol blended in gasohol, a low rate of oil fund levy and an 8-year corporate income tax exemption (Griggers 2004). The government also provides B2 manufacturers a subsidy of 0.30 baht/L from the State Oil Fund to support this program (APEC 2008).

Ajarin Pattanapanchai, the BOI's executive investment adviser said maximum permissible tax incentives have been granted for alternative energy investments since 2004 including an 8-year tax holiday and several non-tax incentives. In 2007, BOI approved 89 projects in the alternative energy sector worth baht 72.8 billion; of these, 40 were for agrofuel plants with a combined value of baht 45.87 billion. Between 2004 and 2006, the BOI approved 26 similar projects worth baht 6.54 billion (Ekvitthayavechnukul 2008).

Retail prices for E85 are 30–40% cheaper than premium gasoline. The government also provides higher tax incentives to automobile manufacturers who invest in compatible E85 vehicles (APEC 2008). Moreover, effective from January 1, 2008, cars using gasoline containing at least 20% of fuel ethanol can get an excise tax reduction. The excise tax cut is expected to lower the price of cars by at least 10,000 baht. The Excise Department estimates that about 30,000 new vehicles powered by E20 or higher will be in the market in 2008 (DEDE 2003) (APEC 2008).

Thailand's Ministry of Energy is cooperating with several ministries including the Ministry of Agriculture and Cooperatives to emphasize R&D to improve production of agriculture products as well as promote use of agrofuels through a model "Biodiesel Production Plan" (DEDE 2003).

Both Indonesia and Malaysia – two countries that control about 85% of global palm oil production – have set up ambitious policies and subsidies for promoting agrofuels. The Indonesian government allocations for agrofuel development between 2006 and June 2008 reached up to US$1.6 billion (although not all of these funds were disbursed). Beginning in 2009, the Indonesian government also pays a subsidy to agrofuel producers to ensure the survival of the industry since agrofuel became more expensive than crude oil-based fuel after oil prices dived more than 70% from their peak in July 2008. Indonesia will only pay the subsidy if agrofuel prices are higher than crude oil-based fuels. Malaysia and Indonesia announced that they would cooperate to stabilize prices for palm oil, which have plummeted since the summer.

Malaysian subsidy support totalled US$16 million in low-interest loans in 2004 and US$3.3 million in federal grants for pilot projects in 2006. Both countries have also set up mandates for blending gasoline and diesel with a certain percentage of ethanol or biodiesel) (Lopez and Laan 2008).

There has been debate in Europe (Doornbusch and Steenblik 2007) but almost none in Thailand over the setting of mandatory production targets and subsidies for agrofuels due to concerns that paying for the agrofuel subsidies could result in an increased tax burden while also further pushing up food prices.

The OECD's Round Table on Sustainable Development report questions the wisdom of setting mandatory targets for agrofuel production when sustainable supply is unknown and commercialisation of second-generation technologies remains uncertain (Doornbusch and Steenblik 2007). The OECD report also argues that agrofuel subsidies reduce the retail price of transport fuels where drivers are insulated from the true costs to society of fuel consumption. The OECD calls for a more efficient policy that would tax fuels according to the externalities they generate and government support for research and development (R&D) into second-generation agrofuels as a more cost-effective way than supporting production from first-generation agrofuel facilities.

Expansion of Feedstock

Ethanol (Cassava and Cane Sugar)

In 2004, Thailand's National Ethanol Committee targeted the production of one million litres (about 264,000 gal)/day of ethanol in the country. The government has approved construction of eight new ethanol plants expected to produce a total of 1.5 million litres (about 396,000 gal)/day (Alternative Transportation Fuels Today, 13 May 2004).

In 2003, Thailand produced 7.2 million tons of sugar of which about two-thirds was exported. The government wants to reduce exports and divert up to 2.6 million tons of sugarcane cultivation per year to ethanol production (Gonsalves 2006). Also in 2003, Thailand exported about 1.3 million tons of molasses; at a conversion rate of 250 L of alcohol per ton of molasses, this represents 328 million litres of ethanol annually.

Originally, sugarcane was used to produce ethanol; however, the production cost of sugarcane is high and the volume produced for ethanol is low since the main purpose of growing sugarcane is for sugar production. So Thailand is looking to increase cassava-based ethanol production (Artachinda 2000).

Thailand is Asia's largest producer of cassava with an average cassava root production of 20 million tons/year. Of this, about 8 million tons are used for domestic starch consumption; another 8 million tons are for making cassava chips for export. The remaining 4 million tons/year can be used for ethanol. At a conversion rate of 6 kg cassava root per litre of ethanol, this will yield 1.8 million litres/day (Gonsalves 2006).

The government plans to boost its current biofuel production in the next 10 years through increasing cassava yield from 23 to 50 t/ha by 2020, and its sugarcane yield from 56 to 106 t/ha by 2021, according to Amnuay Thongsathitya, inspector-general of Thailand's Ministry of Energy (IPS 2008).[4]

Biodiesel (Palm Oil and Jatropha)

Thailand's palm oil economy is third in the world after Indonesia and Malaysia. The country's largest source of raw palm oil is from the southern provinces such as Krabi followed by Surat Thani and Chumphon. Annual crude palm oil output totals 1.3 million tons with about 800,000 t going to the food sector. Of the 500,000 t used in non-food businesses, 420,000 t will be needed to make B2. For B5, at least 600,000 t would be required.

A joint working group from the Ministry of Agriculture and Cooperatives and the Ministry of Energy called "Committee on Biofuel Development and Promotion" (CBDP) has mooted plans to expand the oil palm areas by 640,000 ha in the 5 years from 2008 to yield 4.8 million litres/day of biodiesel (Gonsalves 2006). One key component of the biodiesel strategy suggested by the joint working group is the development of oil palm and jatropha plantations. An estimated total

[4] Breeding and crop improvement research on other important crops such as rice, cassava, and sugarcane (the three principal crops of Thailand) are almost exclusively conducted by the public sector. The Ministry of Agriculture and Agricultural Cooperatives is the largest performer of agricultural research, with an annual research budget of $80 million to $90 million for research on crops, livestock, forestry, and fisheries. Public universities also have significant programs in agricultural research, funded through the Ministry of University Affairs and through grants from the Thailand Research Fund and the National Research Council. A $10-million annual biotechnology research program, most of which is devoted to agriculture, is funded through the National Science and Technology Development Agency, an autonomous public corporation under the Ministry of Science, Technology, and Energy. In addition to investing in public research, the Thai Government has also encouraged private investment in research, although these efforts appear to have had only limited success (TDRI 1990).

investment of 70 billion baht ($1.75 billion) is required for developing the plantations (Gonsalves 2006)

Thailand is also looking at developing 160,000 ha of oil palm plantations in its neighboring countries of Laos, Cambodia and Burma/Myanmar that could add 1.2 million litres/day of additional biodiesel production capacity (Gonsalves 2006).

The committee will also target greater palm productivity from 19 t/ha to 22 t/ha and increase the crushing rate of crude palm oil from 17% to 18.5% by 2012. The government will provide low-interest loans to oil palm farmers participating in these programs (Preechajarn et al. 2008). A few biodiesel plants utilise used cooking oil as feedstock.

Jatropha is also increasingly viewed as an alternative raw material for biodiesel production (APEC 2008). Thailand is planning combined jatropha and oil palm plantations to produce a further 2.5 million litres/day. It is expected that jatropha plantations with an area of 640,000 ha and oil palm plantations of about 192,000 ha will provide 2.5 million litres/day of biodiesel (Gonsalves 2006).

However, the expansion of the oil palm areas has proceeded slowly. In 2006, increased palm acreage was only 48,000 ha, 40% below the annual target. Slower than expected expansion is attributed to more attractive returns from rubber production and the relative lack of incentives for oil palm. To avoid competition with rubber, the government is promoting oil palm plantations in the non-rubber growing areas in the northern and northeastern provinces of Thailand (Gonsalves 2006).

Environmental Consequences

The impacts of agrofuels on the environmental are varied and complex: Impacts may include deforestation, increasing water scarcity, pesticide pollution, soil and water degradation and loss of biodiversity. Much depends on which agrofuel crops are grown, where and how (Rajagopal and Zilberman 2007; Woods Institute for the Environment 2006).

Carbon Costs and Emissions Reductions

The benefits in terms of CO_2 emission reductions and energy security for example vary depending on the type of energy crop, cultivation method, conversion technology and country or region under consideration (Dufey 2006).

The apparent "greenness" of agrofuels was initially the result of focusing on carbon dioxide emissions from the combustion of agrofuels in the vehicle engine which being absorbed from the atmosphere during the growth of the biomass feedstock, was considered "carbon-neutral". The concept of carbon-neutrality has since been revised when considering other emissions of greenhouse gases during cultivation (from production and application of fertilizers, pesticides, herbicides and the use of agricultural machinery) as well as from transportation and processing of intermediates to the final product (Nguyen et al. 2007a, b; Nguyen and Gheewala 2008a, b).

Greenhouse gas (GHG) benefits of agrofuels depend greatly on how they are produced: if forests or peatlands are cleared, there may be a net increase in CO_2 emissions (Fargione et al. 2008). A study published in Science magazine calculated the total carbon costs of agrofuel production and reported that when land clearance (caused either directly or by the displacement of food crops) is taken into account, all the major agrofuels cause a massive increase in emissions (Searchinger et al. 2008). By using a worldwide agricultural model to estimate emissions from land-use change, the study found that for the US, corn-based ethanol, instead of producing a 20% savings, nearly doubles GHG emissions over 30 years and increases greenhouse gases for 167 years (Searchinger et al. 2008).

One study (Scharlemann and Laurance 2008) assessed the total environmental impact of each agrofuel crop by aggregating natural resource depletion and damage to human health and ecosystems into a single indicator, using two different methods. The second key criterion for each fuel is its greenhouse-gas emissions relative to gasoline.

The findings showed that most (21 out of 26) agrofuels reduce greenhouse-gas emissions by more than 30% relative to gasoline. But nearly half (12 out of 26) of the agrofuels – including the economically most important ones such as US corn ethanol, Brazilian sugarcane ethanol and soy diesel, and Malaysian palm oil diesel – have greater aggregate environmental costs than do fossil fuels.

Some studies show Brazilian ethanol in a positive perspective (Goldemberg 2008; Goldemberg et al. 2008; Goldemberg and Guardabassi 2009). Agrofuels that showed best ratings are those produced from residual products such as biowaste or recycled cooking oil as well as ethanol from grass or wood (Scharlemann and Laurance 2008)

Wide variations in greenhouse gas savings from the use of agrofuels appear depending on feedstock, cultivation methods, conversion technologies, and energy efficiency assumptions. For instance, maize-derived bioethanol shows the "worst GHG emission performance" where sometimes the GHG emissions can be even higher than those related to fossil fuels (Peskett et al. 2007). On the other hand, GHG reductions are considered possible from sugarcane-based bioethanol and 'second generation' of agrofuels such as lignocellulosic bioethanol and Fischer-Tropsch biodiesel.

Thailand's initial policy responses on climate change did not incorporate agrofuels (OEPP 2000, p. 81). Subsequent government policy refers to alternative energy and the use of agrofuels; Thailand has also proposed two biodiesel projects for Clean Development Mechanism (CDM) status under the UN Climate Change Convention (Gonsalves 2006).

With the second largest economy in the ASEAN region, Thailand is the second largest in CO_2 emissions in the region; Thailand ranks among the top 30 CO_2 emitters in the world and the seventh largest emitter in Asia. Thailand's per capita emissions of 4.2 t of CO_2 per year is higher than China at 3.8 t per capita, Indonesia at 1.7 t, and India at 1.2 t (UNDP 2007). Thailand's total CO_2 emission was 206 million tons in 2004, which was 2.62 times the corresponding figure in 1990. Thailand's share in the global total CO_2 emission increased from 0.4% in year 1990 to 0.8% in 2004 (Shrestha and Pradhan 2008).

Only a very large agrofuels production cum procurement program, that is far bigger in scale than conceived presently, would help contribute significantly to total national CO_2 emission reduction (Shrestha and Pradhan 2008). However, a massive program for a substantial increase in agrofuels production would face the standard conflict of having to use the agricultural land for fuel production instead of food production. A significantly larger CO_2 reduction potential can also be achieved by shifting the passenger transport demand away from the low occupancy road based personal transport system to urban rail networks like MRTS and railways (Shrestha et al. 2008).

Land Conversion and Deforestation

The type of land converted to agrofuels determines the environmental impacts of the land conversion. Clearing new land creates an agrofuel "carbon debt" as rainforests, peatlands, savannas or grasslands are replaced by crop-based agrofuels in Brazil, Southeast Asia, and the US. This process releases 17–420 times more CO_2 than the annual greenhouse gas (GHG) reductions that these agrofuels would provide from replacing fossil fuels (Fargione et al. 2008). Using fire to clear land for agrofuels such as in China, Indonesia and Brazil can also lead to reducing air quality (Fargione et al. 2008; Peskett et al. 2007). Concerns of ripple effects or leakage (induced change of land use elsewhere as food growing areas are turned into agrofuel) are also being discussed (Kløverpris et al. 2008; Schmidt 2008).

The time required for agrofuels to overcome this carbon debt due to land use change and begin providing cumulative greenhouse gas benefits is referred to as the "payback period"; this has been estimated to be 100–1,000 years depending on the specific ecosystem involved in the land use change event (Kim et al. 2009).

Recent research points out that land use change studies have not yet addressed several variables like cropping management that is a key factor in estimating greenhouse gas emissions associated with land use change viz. the effects of different tillage methods on land use change (Kim et al. 2009). If management practices are poor, it may also have substantial effects on other ecosystem services (Robertson et al. 2008).

In Brazil, the country most famous as a low-cost ethanol producer, the agrofuels development plan "Avança Brasil" – for which the government allocated about US$ 40 billion – seeks to expand the agricultural frontier deep into the forests in the north of the country that could result in widespread deforestation (WRM 2007).

In Indonesia, Wilmar, the world's biggest trader in palm oil, is allegedly illegally logging rainforests, setting forests on fire and violating the rights of local communities in Indonesia, according to a report by Friends of the Earth International (FOE 2007). The Singapore-based multinational Wilmar claims they are only developing plantations on land that is approved by the Indonesian government for the cultivation of oil palm. Wilmar is a member of the industry-led Round Table on Sustainable Palm Oil (RSPO) and is funded by the World Bank's private arm as well as private European banks, which have codes of conduct against unsustainable palm oil.

Clearing land for oil palm plantations in Indonesia has led to numerous conflicts with local communities. The chair of the UN Permanent Forum on Indigenous Issues has warned that five million indigenous people in West Kalimantan alone are likely to become refugees because of biofuel expansion (CSR 2007).

A report about the impacts of palm oil production published by Friends of the Earth found that between 1985 and 2000, the development of oil palm plantations was responsible for an estimated 87% of deforestation in Malaysia. In Sumatra and Borneo, some 4 million hectares of forest has been converted to oil palm farms (FOE 2005).

In Thailand, the current production of agrofuels feedstock has not been associated with recent forest conversion. Some plantations of sugarcane and cassava (used for ethanol) and oil palm (used for biodiesel) were already in existence long before the promotion of agrofuels. There have been some shifts among crops based on real or perceived economic benefits. Although the Thai Ministry of Agriculture and Cooperatives policy is that new plantations are to be on abandoned rice fields, deserted public lands and degraded areas, in practice there are concerns that expansion of oil palm plantations may lead to forest encroachment (Bangkok Post 2008a; Siriwardhana et al. 2008). As a climate change mitigation strategy reducing deforestation is better than converting forest to agrofuel production (Danielsen et al. 2008).

Water-Use

Agrofuels lead to high water consumption in two areas: the esterification process with methanol and the production of chemical fertilizers (Puppán 2002). Large-scale agrofuel use could alter the distribution of water locally, regionally, and globally (Woods Institute for the Environment 2006). Concerns are that in regions already under water stress, biofuel production may further decrease the freshwater availability for other uses and for meeting peoples' basic needs (Varghese 2007).

The International Water Management Institute (IWMI) estimates that if current global plans to produce biofuels go ahead, 180 km^3 of additional irrigation water will be needed. It says that "while in some areas this won't put too much stress on water supplies the more ambitious plans in China and India to boost domestic production of agrofuels raise serious concerns for future water supplies if traditional food crops – maize in China and sugarcane in India – are used" (IWMI 2008a).

One modelling study with a special focus on India and China explored the land and water implications of increased agrofuel production. The study concluded that local and regional impacts on water from agrofuel crops could be substantial (Fraiture et al. 2008). The study warned that the strain on water resources in China and India could be so much that policy makers could decide to not pursue agrofuel choices based on traditional field crops.

One litre of agrofuel production requires approximately between 1,000–3,500+ L of water to grow the biomass required. IWMI for example estimates that "it takes on average roughly 2,500 litres of crop transpiration and 820 litres of irrigation water withdrawn to produce one litre of biofuel" (IWMI 2008b). But the variation within regions can be quite large (Uhlenbrook 2008; WWAP 2009). In Europe, where rain-fed rapeseed or cereal is used, irrigation requirements for agrofuel crops are negligible. In the US, places with rain-fed maize use only 3% of all irrigation withdrawals for agrofuel crop production while 400 L of irrigation water are required per litre of ethanol. In China or India, irrigation of maize or sugarcane requires up to 3,500 L of water withdrawal per litre of agrofuels (EuropaBio 2008).

The impacts on water depend on the crop and the scale at which it is grown. Sugarcane especially requires large volumes of water that can pose problems in semi-arid areas (Peskett et al. 2007). Existing water allocation could come under stress when corn croplands expand into areas requiring irrigation together with the installation of water-intensive bio-refineries in areas of water scarcity creating problems in areas without a history of water quantity problems (Woods Institute for the Environment 2006).

Agrofuel facilities require water to convert the biomass into fuel. While water use in the bio-refining process may be modest, since the water consumption is concentrated into a smaller area, the impacts to the local area are multiplied (Biswas and Seetharam 2008).

Land use for agrofuels can affect water availability, plant nutrition and soil quality (Zuurbier 2008). Depending on the scale and type of crops are grown, agrofuels can pose a wide range of impacts: accelerate existing levels of topsoil erosion, deplete soil carbon and nutrients, compact soils, deplete groundwater and increase water pollution, as well as loss of biodiversity (Tegtmeier et al. 2004). But the picture is not as clear-cut since these issues are as true for agriculture for agrofuels or food.

In Malaysia's palm oil extraction, about 1.5 t of palm oil mill effluent (POME) – that contains acid and has a high organic load – is produced per ton of fresh fruit bunch (FFB) processed by the mill (Ahmad et al., 2005). POME in Thailand's oil palm processing areas is known to significantly contribute to surface water pollution (Setiadi and Husaini 1996). Apart from soil and water contamination, toxicity to workers is of concern from the use of herbicides such as Paraquat and Glyphosate in oil palm plantations; the insecticide Furadan is applied in the oil palm nursery (Pleanjai et al. 2004). Furadan is the brand name of the pesticide Carbofuran that has faced controversy since the 1980s after the US Environmental Protection Agency (EPA) Special Review estimated that over a million birds were killed each year by the granular formulation.

The new generation of cellulosic feedstock ethanol made from cellulose-rich organic material expected to be commercially viable within the next 5–15 years (Woods Institute for the Environment 2006) will shift ethanol feedstock away from corn kernels to agriculture and forestry residue, grasses and woody crops, and the organic portion of industrial and municipal solid waste. Native grasses and perennials are less water-, pesticide-, and fertilizer-intensive. However, water quality improvements are not guaranteed as different cellulosic feedstocks have dramatically different effects on water quality (Woods Institute for the Environment 2006).

In northeast Thailand, where cassava and sugar cane are the major raw materials for the production of ethanol (and jatropha plantations are also being planned), drought conditions can pose uncertainties for agrofuels production. Northeast Thailand is frequently subject to drought due to erratic rainfall distribution patterns, critical dry periods within the rainy season, and low soil water holding capacities (Siripon et al. 2000). Successive Thai governments have planned huge investments for large-scale irrigation projects and intra-basin transfer of water from the Mekong River to expand irrigation in the northeast region, viewed as the most arid region in the country (Molle et al. 2009). The overall context of contested water resources poses complex governance challenges for this key region now being targeted for agrofuels expansion.

Economic Considerations

Competitiveness

Agrofuels are expected to reduce dependency on foreign oil imports, reducing trade deficits, and improving energy security (Hoekman 2009). In Thailand, the competitiveness of the ethanol price relative to gasoline is crucial since the Thai government's plan is to substitute ethanol in octane 95. The ethanol prices in turn are highly dependent on the feed stock prices as volatility of the feedstock prices in the market significantly affects the competitiveness of the ethanol relative to gasoline.

The degree to which agrofuels can fully displace fossil fuel energy varies between different estimates, technologies and countries. The biggest commercial barrier to agrofuels is their economic competitiveness compared to fossil fuels.

For ethanol (blended with gasoline) from both cassava as well as cane molasses, though the environmental costs were lower than that of gasoline, the overall cost was still higher. Studies on the GHG abatement cost show that with oil prices in late 2006 or early 2007, ethanol from neither cassava nor sugarcane (molasses) is competitive without government intervention (Nguyen et al. 2007a, b). According to Thailand's Energy Planning and Policy Office, cassava-based ethanol is more costly to produce, at 21–22 baht/L, compared to Thailand's benchmark ethanol price of 18.01 baht/L (Praiwan 2008).

Calculating the GHG abatement costs on the basis of the reduction in GHG emissions and the price difference between ethanol and gasoline showed values of about US$100/t of CO_2-eq; there are far more cost-effective ways of reducing GHG emissions than the use of 10% ethanol blends. It must be noted here that land use change has not been included.

The economic feasibility of biodiesel is also under question due to the high feedstock prices. Biodiesel is more expensive than conventional diesel. At 2008 costs, it becomes competitive when crude oil prices are around US$100–120 (Siriwardhana et al. 2008). Product cost is strongly determined by cost of crude palm oil and to a lesser extent methanol (Siriwardhana et al. 2008). Biodiesel could become more competitive if yields of oil palm could be improved and costs of the key input thus brought down (Kapilkarn and Peugtong 2007). However, as in the case of ethanol, studies must also be done to internalize the environmental benefits of replacing diesel with palm biodiesel.

The government has to ensure feedstocks are available throughout the year with competitive prices. Feedstock availability varies from season to season and with geographic locations; their prices are also volatile, which in turn affects ethanol production costs. Nearly all of Thailand's existing agrofuel plants faced supply surpluses as well as increased input prices by mid-2008; nearly all ethanol plants were running at only 70% of their production capacity while some either suspended production or switched to non-ethanol products (Preechajarn et al. 2008).

One solution is to use many different types of feedstocks for the ethanol production. It is mooted that the crops like sugarcane and cassava, due to their low average prices as feed stocks, are most likely to be able to compete with gasoline (Nguyen and Gheewala 2008b).

Under What Conditions Are Agrofuels Competitive?

Ethanol

The overall cost of the ethanol blends (with gasoline) from both cassava as well as cane molasses, could be reduced below that of gasoline by substituting the fossil energy used in the ethanol conversion stage by biomass-based fuels for process energy. The competitiveness of cane molasses based ethanol could further be improved by a change in the current agricultural practice of open burning the cane trash, which could instead be used, as an energy source.

Also to be noted are the savings in gasoline due to the substitution by biomass-based ethanol produced from local feedstock resulting in reduced oil imports and also having a positive effect on energy security. The ethanol blends could be even more competitive if these could be monetized and included in the overall cost calculations.

A more detailed life cycle cost analysis revealed that about two thirds of the life cycle cost of cassava ethanol is contributed by the feedstock (Nguyen et al. 2008). Cassava ethanol can be made more competitive by reducing the feedstock cost by an increase of yield (from 0.087 to 0.118 ha^{-1} which is technically feasible) through appropriate farming practices without compromising the farmers' incomes.

Also, the cost competitiveness can be further improved at the ethanol conversion stage by utilization of by-products (stillage sludge as manure and CO_2 for soda manufacturing). Adding the external benefits from reduced fossil oil use and GHG reductions makes the cassava ethanol price competitive with gasoline. The study showed that the increase of cassava yield in fact would result in increased farmer incomes as well as more opportunities for employment. Scientists from the Centre for Tropical Agriculture (CIAT) are working to develop new cassava varieties to enable smallholder farmers in Thailand to produce more revenue. CIAT's 5-year collaboration with Bangkok-based Thai Tapioca Development Institute (TTDI) will attempt "to adapt cassava to growing conditions in Thailand while optimizing its marketability" (CIAT 2008).

The above studies indicate that the government subsidies for promoting ethanol (see below and Section Tax Breaks and Subsidies) could to some extent be justified. However, highly reduced prices would encourage excessive fuel usage thus promoting non-sustainable consumption.

Biodiesel

Studies in Thailand have shown that using biodiesel can reduce the GHG emissions by more than 75% as compared to diesel provided of course that tropical rainforests or peatlands are not being destroyed to plant oil palm (Pleanjai et al. 2009a). This will be particularly critical when the expansion plans for oil palm are considered (c.f. Section Expansion of Feedstock).

Also, there are opportunities for improving the utilization of by-products that would further increase competitiveness. A smaller, but currently utilized feedstock for biodiesel is used cooking oil for which the GHG benefits from utilizing this resource are substantial (Pleanjai et al. 2009b). The projected savings on diesel imports accruing from a theoretical utilization of all the more than 100 million litres of used cooking oil has been estimated as US$40 million (Siriwardhana et al. 2008) Added to these are the benefits of diverting used cooking oil from being reused.

For another feedstock for biodiesel, jatropha, competitiveness at the commercial scale is not yet ascertained. Preliminary studies with the life cycle energy balance of jatropha biodiesel have shown encouraging results (Prueksakorn and Gheewala 2008). Further studies are required to establish the conditions under which adequate benefits could be obtained from the point of view of economic and social sustainability.

Social Assumptions and Implications

The expansion of agrofuels raises issues of social justice or fairness over the allocation of benefits, burdens and risks. Apart from questions of who gets to produce, sell and distribute energy that we dealt with above, these can be largely distilled into issues of food security, rural livelihoods and consumer benefits.

Food Security

After almost a decade of low or stagnant agricultural commodity prices, the last couple of years have seen rising food prices. In December 2007, the Food and Agriculture Organization (FAO) calculated that world food prices rose 40% in the previous 12 months and affected all major agrofuel feedstocks: sugarcane, corn, rapeseed oil, palm oil, and soybeans. On 17 December 2007, the FAO head Jacques Diouf warned that there is "a very serious risk that fewer people will be able to get food," particularly in the developing world (Tenenbaum 2008).

While the rapid price increases cannot be blamed solely on agrofuels, the demand for agrofuels is seen as overwhelming an already overextended food supply system. Agrofuels were also identified as a major culprit by the UN, World Bank, and International Monetary Fund (IMF). The IMF estimates that in 2007 they accounted for almost half of the increase in demand for major food crops (IMF April 2008).

Moreover, the agrofuels demand also affects non-feedstock crops such as rice and wheat as farmers plant feedstock instead of food (Tenenbaum 2008). By OECD estimates, between 2005 and 2007, almost 60% of the increase in consumption of cereals and vegetable oils was due to agrofuels (OECD 2008).

Agrofuels compete indirectly with food – for land, water, and other inputs – further pushing up food prices. The International Food Policy Research Institute (IFPRI) is calling it a "tax on food' that may have severe effects felt most by poor people as support for agrofuels is an incentive for the diversion of crops and agricultural land away from food production into fuel production (IFPRI 2007).

In Thailand diversions of molasses, sugar and oil palm to agrofuels may make basic alternative food products more expensive. Palm oil, for example, is a key cooking oil in Thailand comprising about 60–70% of market share (Siriwardhana et al. 2008). Prices, globally, have risen as palm oil is also used increasingly and widely in processed foods. Jatropha, as a non-edible plant oil, may be a suitable alternative to oil palm, in the sense of reducing the direct competition between food and fuel uses (Siriwardhana et al. 2008)

The case for biodiesel in Thailand is different from that of ethanol in that there is a direct conflict between food and fuel as the biggest feedstock for biodiesel, palm oil, is also very popular as cooking oil. Recent increases in palm oil prices have been partly blamed on its expanded use for biodiesel. In fact, due to shortages in supply, Thailand recently had to import 30,000 t of crude olein for cooking purposes (Bangkok Post 2008a). Plans to expand oil palm plantations should be very carefully studied to avoid replacing food farms (fruit, coffee and rice fields) (Bangkok Post 2008b, c).

Ripple Effect: Energy and Food

The long-term implications for food security appear to revolve around the now increasing integration of the energy and agricultural sectors, and the consequent interplay among the agrofuels markets, food commodities and petroleum prices (Naylor et al. 2007).

As food crops can now be converted into fuels, a new factor comes into play: the link between the price of food and the price of petroleum. With the construction of so many fuel ethanol distilleries, the food and energy economies that have largely been separate are now merging (Tenenbaum 2008).

The influence of energy prices on agricultural markets is now a crucial factor. Traditionally, the transmission of energy prices to food markets was in terms of energy *inputs* such as fertilizer, mechanization and transportation. The rapid growth in agrofuels production and markets is putting pressure on the main agricultural commodities that have long been used for food and animal feed. This is leading to escalating prices in the international markets not only of these main feedstocks but also other substitutions causing price increases in a range of agricultural markets. The second is the price increases are also linked to the petroleum prices: as long as petroleum prices remain above US$55–60 per barrel, the demand for agrofuels and food price increase seem set to continue (Naylor et al. 2007).

The continuing expansion plans for agrofuels production capacity and plantations in several countries seem set to generate 'ripple effects'. According to Naylor, "The ripple effects will be either positive or negative if energy markets begin to determine the value of agricultural commodities and the long-term trend of declining real prices for most agricultural commodities could be reversed. Over the short term, this reversal, while helping net food producers in poor areas, could have substantial consequences for the world's food-insecure, especially those who consume foods that are direct or indirect substitutes for agrofuels feedstocks. Food price volatility has the largest impact on extremely poor households, who typically spend 55–75% of their income on food" (Naylor et al. 2007).

Jean Ziegler, the UN's special rapporteur on the right to food, called for a 5-year international ban on producing agrofuels in order to combat soaring food prices. Ziegler said that the conversion of arable land for plants used for green fuel had led to an explosion of agricultural prices that was punishing poor countries forced to import their food at a greater cost (Rowe 2008).

Prices of feedstock like cassava are also pushed by China's demand over the past few years as China has emerged as the leading cassava importer, procuring mostly feed ingredients. Presently, the country accounts for around 60% of the global market.

Growth in China's imports of cassava has been due partly to implementation of a free trade area between Thailand and China which led to the abolition of a 6% tariff on cassava products, but the main driver of growth is China's hugely expanding livestock industry(Prakash 2008). The rising demand for cassava as animal feed particularly in southern China will make it difficult for Thailand's fuel producers to source cheaper cassava for fuel. China's feedstock imports for biofuel production are expected to rise over the coming years to meet its demand for fuel ethanol production (Latner et al. 2006).

In response to food security concerns, Thailand's Office of Agricultural Economics (OAE) has been entrusted with working out the policy of food security, According to the OAE, "the policy of food security seeks to increase the efficiency of food production. It will ensure that the planting of energy crops will not have adverse effects on food production. The policy will also add value to agricultural products, protect farmland, and set zoning for the planting areas of certain crops" (Foreign Office Thailand 2008).

Consumers – Energy Users

Thailand's companies and users of palm oil are seeking to produce agrofuels which comply with sustainability standards and have sought help from the Thai government to promote energy and "eco-efficiency" with the "E3-Agro Project". Thailand is starting a certification scheme for oil palm in cooperation with Germany's Federal Environment Ministry in a 3-year US$4.7 million project. The project aims to encourage "sustainable production of palm oil for biofuel" and promote the supply of sustainable palm oil to Europe and establish a certification scheme for sustainably produced palm oil in Thailand. The Thai-German project is undertaking a study to look at the performance of selected Thai palm oil mills in comparison to international sustainability standards (SEA-CR 2007). The standards focus on sustainable production of biomass for agrofuels covering ecological, economical, social, as well as political issues, as well as on the greenhouse gas reduction potential. The study will also include special focus on sustainable cultivation of land, protection of natural habitats and give special attention to the needs of smallholders (SEA-CR 2007).

The pilot programme which will start in Krabi will support the introduction of the Roundtable on Sustainable Palm Oil (RSPO) standards (Bioenergy Business 04 March (2009). The RSPO, created in April 2004, is an association that includes oil palm growers, banks and nongovernmental organisations "to promote the growth and use of sustainable palm oil through co-operation within the supply chain and open dialogue with its stakeholders."

The EU is already active in the development of certifying agrofuels and biomass; on 23 January 2008, the European Commission introduced the draft directive on the promotion of the use of energy from renewable sources which includes, among many other provisions, a set of "agrofuels sustainability criteria" to be applied to domestic production and imports to reach the EU's 10% target by 2020 (Naylor et al. 2007).

Efforts at certification are underway with some food and consumer goods companies wanting a plan to certify palm oil to ensure that production doesn't result in destruction of tropical rainforests. The push for the so-called "green palm oil" is joined among others by Unilever, Johnson & Johnson, Nestlé SA and H.J. Heinz Co. who have signed up with a consortium of 200 oil producers, commercial buyers and environmental groups to improve the industry's image and avert a consumer backlash (Wright 2008).

Consumer concerns about agrofuels environmental and energy security issues as well as usability in standard automobile engines. Consumer confidence in agrofuels increased once the Thai government secured agreement with car manufacturers to maintain warranties on existing and new diesel vehicles if they use 5% BD fuel (Siriwardhana et al. 2008).

Consumer issues of compatibility and quality control remain important for users who expect equivalent (or better) performance and convenience. Low ethanol–gasoline blends (5–10%, E5–E10) can fuel gasoline vehicles with little if any engine modification. But consumer concerns persist in particular about ethanol drawbacks such as compatibility with some plastics or metals (Al-alloys, brass, zinc, lead) and high latent vaporization heat (making engines hard to start).

Presently, Thailand's service stations sales of E10 gasohol account for about 20% of total petroleum sales (APEC 2008). Following a surge in E10 gasohol sales in 2007, as of 1 January 2008, the government launched E20 gasohol (Preechajarn et al. 2008) and the Petroleum Authority of Thailand (PTT) and Bangchak Companies started supplying E20 since then. Bangchak has introduced E85 at its stations (APEC 2008).

For biodiesel, B2 is already available nationwide; PTT and Bangchak started selling B5 in 2007 (APEC 2008). The supply of B5 is expected to increase in 2007 and B10 in 2012. There are smaller-scale community level biodiesel plants with capacities from 5,000 to 20,000 L/day that serve communities in 11 provinces (Gonsalves 2006).

While the price of fossil fuels remains relatively cheap, low cost agrofuels for consumers can only be competitive with government subsidies. Biodiesel fuel mix (B20) produces less particulate matter than standard diesel and so could contribute to efforts to improve air quality in Bangkok (Siriwardhana et al. 2008).

Rural Livelihoods

The FAO states that agrofuels "offer the opportunity for agricultural and rural development - if appropriate policies and investments are put in place" (FAO 2008). Since its production requires fewer investments and less complex processing plants compared to oil, gas or coal, agrofuels could be within the scope and affordability of Small and Medium Enterprises (SMEs). Rural biofuel processing activities could produce fuel for local use in farm machinery and other smaller engines (Gonzales 2008; Phalakornkulea et al. 2009).

Local agrofuels development could be a source of energy for rural (especially remote) communities such as in small-scale rural electrification. In Tanzania, a non-governmental development organization introduced generators that use SVO (straight vegetable oil) in rural areas using jatropha. The generators can provide power for machinery, recharge batteries and bring electricity to village shops and to households for some hours at night (AU Monitor 2008).

Competition for land with other crops as noted above may lead to higher prices of other commodities for consumers as land is switched to oil palm. Such switches may also benefit wealthier farmer as it involves larger parcels of land. Some proposed policies (Siriwardhana et al. 2008), like preferential soft loans for development of oil palm plantations, could have important side-effects on access and ownership of land for poorer households unable to participate.

Questions remain over how agricultural development patterns will respond to the growth of the agrofuels market and whether benefits will flow to poorer rural households. An emerging issue is also Thailand's plans to invest in agrofuel plantations in neighboring countries such as Burma/Myanmar, Laos and Cambodia and the implications for small-scale food producers.

Food scarcity is a serious problem in many parts of Burma/Myanmar, according to the United Nations World Food Program; in 2007, some 5 million people or almost 10% of Burma's population were chronically short of food. However, the country's military rulers have embarked on a massive expansion of 3.2 million hectares of jatropha plantations for agrofuels where planting quotas are enforced through forced confiscation of lands as well as arrests, fines, and beatings of farmers (ECDF 2008). In Laos, the state and private sector have introduced some biofuel production programs using jatropha. One concern is that farmers have little experience with non-food crops. Also, there are no factories to process jatropha or other agrofuels (Sengmany 2007).

The ability for farmers to earn money selling agrofuel feedstocks "sounds very good on the surface," says Liane Schalatek, associate director of the Heinrich Boll Foundation-North America. When agrofuels replace existing production patterns of small farmers with large-scale monoculture plantations, the people who used farm their lands are turned into seasonal or temporary laborers (Tenenbaum 2008) resulting in loss of livelihood security.

Concerns surround the impacts of agrofuels expansion on women. A recent FAO report looks into agrofuel production and their gendered impacts, explaining that it may increase the marginalization of women in rural areas, threatening their livelihoods (Rossi and Lambrou 2008).

Feedstock price is another critical issue to the profitability of agrofuels (Yoosin and Sorapipatana 2007). On the one hand technical and management innovations may reduce these costs; on the other, parties with interests in agrofuels will likely keep pressure on government to keep input prices low. For instance, the promotion of agrofuels in the US has been influenced by commercial and farming interests (Magdoff 2008).

In the past most agrofuels made have been consumed domestically. In the future this is likely to change as developed countries exploring import options (Lund 2009). There are already some signs of such a transformation arising in response to domestic and regional policies, like the EC Directive on agrofuels (European Commission 2003), and the International Agrofuels Forum (Mathews 2007). If trade in agrofuels was to increase a lot, it may get harder for the Thai government to influence land- and water-uses related to agrofuels than is the case currently. While waste biomass maybe adequate when demand is small, as industry grows dedicated cultivation of feedstock will be needed (Hoekman 2009).

Prospects

Sustainability Questions

Agrofuel sustainability connects many environmental, economic, and social facets where the tradeoffs vary widely depending on the types of fuels and where they are grown (Robertson et al. 2008).

Several concerns have arisen with the first generation agrofuels like biodiesel and gasohol. First, agrofuels may not be a cost-effective way to reduce greenhouse gas emissions. Sometimes it appears they may even use more energy to produce than went into their creation with pursuit only plausible because of subsidies (Eaves and Eaves 2007). Second, there may not be enough land available for energy crops; Thailand may have to use large areas of agricultural land to meet its agrofuel targets. An important issue is the scale of land-use required for achieving the agrofuels planting targets. Agrofuels may work on a low or local level, but meeting the demands of the transport sector is of a different order of magnitude. Agrofuels may offer benefits on a small scale such as for small-scale rural energy use; however, the same is not always true on a large scale (Rowe 2008). Third, crops for agrofuel may compete with food production (Naylor et al. 2007). Fourth, they may be a driver of deforestation (Butler and Laurance 2008; Sharlemann and Laurance 2008). Fifth, they may have environmental effects such as toxic contamination of soil and water as well as farm workers from the chemical herbicides and fertilizers used in the plantations (Galloway et al. 2008; Gordon et al. 2008).

Agrofuels could be part of Thailand's renewable energy mix. In 2003, renewables including solar, wind, micro-hydro and biomass accounted for less than 1% of energy use in Thailand; there appears room for substantial expansion of renewable energy sources. But uncertainties remain about how important a role liquid agrofuels – either current or next-generation – can realistically play in the renewable energy economy. While agrofuels could be an energy source for the future, the most optimistic studies still seem to estimate that by 2050 only about 13% of liquid fuels could be supplanted by agrofuels. In the transport and energy sector, the question then is whether it is justifiable that agrofuels get so much public funding and subsidies while not giving other energy and transport technologies a chance to develop or compete (Doornbusch and Steenblik 2007).

This concern is reflected in a report published in April 2009 of the US Scientific Committee on Problems of the Environment (SCOPE) which views few opportunities in agrofuel production for maximizing social benefits while minimizing environmental impacts, and that even those opportunities that exist are not likely to provide a significant contribution to society's energy budget (SCOPE 2008). Burning biomass itself as fuel to generate electricity and heat is likely a more efficient way to produce renewable energy than converting biomass to liquid fuel such as ethanol (SCOPE 2008).

One reason to pursue agrofuels even when they are not competitive today is to be prepared for when they are. But this can also turn into a self-serving discourse.

Second generation agrofuels based on ligno-cellulose degradation are still a decade or more from large-scale commercialization (Himmell 2007).

Are Targets and Subsidies Justified?

The Thai government has set targets and provided tax breaks and subsidies to bring down the price of bioethanol blends at the pump (see more details in Section Tax Breaks and Subsidies). However, overall decisions on targets appear to be made on little evidence and in the absence of strategic long-term planning and investments. Targets and subsidy mechanisms have often been used to justify promotion of new technologies for fuels in the market with the hope that once they mature, they would become cost competitive. But globally the trend is on the reverse as targets are being scaled down and subsidies diminished or scrapped. Australia has made relatively modest targets mandatory rather than aspirational (Kent and Mercer 2006). The OECD concluded that governments should scrap mandatory agrofuel targets (Doornbusch and Steenblik 2007; Rowe 2008). At a summit in March 2007, EU heads of state added two important conditions to their mandatory agrofuels targets: that agrofuels should be produced sustainably and that second-generation agrofuels should become commercially available (Rowe 2008). The Netherlands government has abolished all subsidies for agrofuels produced from palm oil in November 2006 (Naylor et al. 2007).

One justification could be that the environmental and social benefits provided by agrofuels are not included in the cost of the agrofuels at the pump. Based on our analysis here on external environmental and social benefits, government subsidies are not justified. Fuel subsidies increase consumption, discourage more efficient use of resources and absorb national budgets that could be spent on social services (such as health and education) as well as renewable energy or public transport options.

Policy-Making Challenges

Thailand's administrative restructuring in 2002 reduced the number of agencies and ministries involved in energy policy subsuming most under a new Ministry of Energy (Uddin et al. 2006). The Department of Alternative Energy Development and Efficiency (DEDE) took over primary responsibility for renewables with the Energy Planning and Policy Office (EPPO) involved in higher-level analysis and strategy development. Nevertheless fragmentation problems persist as goals in different ministries still diverge (Prasertsan and Sajjakulnukit 2006). A plethora of committees that are often disconnected and short-lived appear to be part of the policy-making problem.

Governments and other actors need to be clear on why they are promoting agrofuels in the first place and the basis for expectations and targets. Moreover, there needs to be greater transparency in decision-making and separation of government politicians and collusion with private sector "interests". Thailand's Ministry of Energy, for instance, often has Petroleum Authority of Thailand (PTT) executives in key positions.

One of the recurrent challenges in Thai policy-making context has been to move from targets and budgets to achievements and expenditures. Experts in Thailand have argued that current financial incentives for biomass energy are insufficient and other institutional, policy and information barriers are still substantial (Prasertsan and Sajjakulnukit 2006). They agree that the wider public is not yet convinced and acknowledge that labour unions are concerned about job security. Addressing these barriers in a timely way will require much greater participation of stakeholders than has so far been typical in energy policy-making in Thailand, in general, and with respect to renewables, in particular.

In summary, the market costs of agrofuels in Thailand seem at present to be not competitive with fossil fuels without government support. Given external environmental and social concerns, our analysis shows that government subsidies are not justifiable. There are many opportunities for improving the efficiency of production and by-product utilization, which would make them more favourable in terms of economic competitiveness. However, if environmental and social benefits are considered, then the picture becomes different. Thailand would be better served with policies focused on reducing energy consumption, improving energy efficiency and influencing changes in societal patterns of production and consumption.

References

Ahmad AI, Ismail SB, Bhatia S (2005) Water recycling from palm oil mill effluent (POME) using membrane technology. J Desal 157(1–3):87–95

APEC (2008) Thailand biofuels activities. APEC Biofuels

ARDA (2005) Bio-diesel and Thai Community Development. In: Proceedings of the first conference on plants for energy research, Agricultural Research Development Agency (Public Organisation) (ARDA), Bangkok

Artachinda A (2000) Bio fuel development and consumption in Thailand. Mekong Environment and Resource Institute (MERI), Bangkok

African Union (AU) Monitor (2008) Biofuel and Food Security. African Union (AU) Monitor

Bangkok Post (2008a) Bio-diesel squeezes palm supply. January 16, 2008.

Bangkok Post (2008b) Should Thai cars be gorging themselves on Thai food? January 16. Available at: http://www.readbangkokpost.com/business/oil_and_energy/should_cars_eat_our_food_14010.php

Bangkok Post (2008c) Will palm oil profits push Thai fruit and coffee off the field? 12 March. Available at: http://www.readbangkokpost.com/business/agriculture/will_palm_oil_profitspush_thai.php

Bioenergy Business (04 March 2009) Germany funds €3.5m Thai sustainable palm oil project

Biswas AK, Seetharam KE (2008) Achieving water security for Asia. Water Resour Dev 24(1):145–176
Butler R, Laurance W (2008) New strategies for conserving tropical forests. Trends Ecol Evol 23:469–472
CIAT (2008) Partnership with TTDI to enhance cassava production and farmer livelihoods in Thailand. International Centre for Tropical Agriculture (CIAT)
CSR (2007) Biofuel refugees. CSR Asia Weekly, vol 5 Week 17 Corporate Social Responsibility Asia
Danielsen F, Buekema H, Burgess N, Parish F, Bruhl C, Donald P, Murdiyarso D, Phalan B, Reijnders L, Struebig M, Fitzherbert E (2009) Biofuel plantations on forested lands: double jeopardy for biodiversity and climate. Conservation Biology 23(2)
DEDE (2003) Nayobay Palangan Thaen Dan Cheuapleung Ethanol. Department of Alternative Energy Development and Efficiency (DEDE), Ministry of Energy, Thailand, Bangkok
Doornbusch R, Steenblik R (2007) Biofuels: is the cure worse than the disease? In: Honnens H (ed) OECD's round table on sustainable development. OECD
Dufey A (2006) Biofuels production, trade and sustainable development: emerging issues. Sustainable Markets Discussion Paper Number 2. Environmental Economics Programme/ Sustainable Markets Group, International Institute for Environment and Development (IIED)
Tegtmeier E et al. (2004) External costs of agricultural production in the United States. Int J Agric Sustain 2(1): 1–20
Eaves J, Eaves S (2007) Neither renewable nor reliable. Regulation Fall:24–27
Ekvitthayavechnukul C (2008) Biofuel agency to be formed: formulating policy for biodiesel and ethanol to be among main remits, The Nation, Bangkok
Ethnic Community Development Forum (ECDF) (2007) "Biofuel by Decree". Chiang Mai, Thailand
EuropaBio (2008) Fact sheet: biofuels and land use. The European Association for Bioindustries. 17 April
European Commission (2003) Directive 2003/30/EC of the European Parliament and of the Council on the promotion of the use of biofuels or other renewable fuels for transport. Official journal of the European communities, 17 May 2003, L123/42–46. European Commission
FAO (2008) Biofuels: prospects, risks and opportunities The State of Food and Agriculture 2008. Food and Agriculture Organization (FAO), Rome
Fargione J, Hill J, Tilman D, Polasky S, Hawthorne P (2008) Land clearing and the biofuel carbon Debt. Science 319:1235–1238
FOE (2007) 'Policy, practice, pride and prejudice' Friends of the Earth (FOE), Netherlands, Amsterdam
Foreign Office Thailand (2008) The Policy of Thai Food Security and Food Safety Foreign Office. The Government Public Relations Department, Bangkok
Fraiture Cd, Giordano M, Liao Y (2008) Biofuels and implications for agricultural water use: blue impacts of green energy. Water Policy 10:67–81
Fungtammasan B (2008) Thailand's renewable energy policies and programs: focus on bioenergy and biofuels. Energy and Environment Office, London
Galloway J et al (2008) Transformation of the nitrogen cycle: recent trends, questions and potential solutions. Science 320:889–892
Goldemberg J (2008) Opinion: the challenge of biofuels. Energy Environ Sci 1:523–525
Goldemberg J, Coelho ST, Guardabassi P (2008) The sustainability of ethanol production from sugarcane. Energy Policy 36(6):2086–2097
Goldemberg J, Guardabassi P (2009) Are biofuels a feasible option? Energy Policy 37(1):10–14
Gonsalves JB (2006) An Assessment of the biofuels industry in Thailand. United Nations Conference on Trade and Development (UNCTAD), Geneva
Gonzales ADC (2008) Overall stocktaking of biofuel development in Asia-Pacific: benefits and challenges. United Nations Economic Commission for Asia and the Pacific (UNESCAP), Bangkok
Gordon LJ, Peterson GD, Bennett EM (2008) Agricultural modifications of hydrological flows create ecological surprises. Trends Ecol Evol 23:211–219
Griggers C (2004) Biofuels: a natural solution to rising oil costs. Thailand Investment Review-Board of Investment (BOI)

Himmell ME (2007) Biomass recalcitrance: engineering plants and enzymes for biofuels production. Science 315:804–7

Hoekman S (2009) Biofuels in the U.S. – challenges and opportunities. Renewable Energy 34:14–22

IFPRI (2007) The world food situation: new driving forces and required actions. International Food Policy Research Institute (IFPRI), Washington, DC

IWMI (2008a) Biofuels and implications for agricultural water use: blue impacts of green energy. International Water Management Institute (IWMI), Colombo, Sri Lanka http://www.iwmi.cgiar.org/EWMA/files/papers/Biofuels%20-%20Charlotte.pdf

IWMI (2008b) Water implications of biofuel crops: Understanding trade offs and identifying options. Water Policy Brief: Putting Research Knowledge into Action. International Water Management Institute (IWMI), Colombo, Sri Lanka

International Monetary Fund (IMF) (April 2008) World economic outlook. International Monetary Fund (IMF), Washington DC

IJO (2007) Jatropha curcas – A country-by-country lookout. Jatropha News Network. International Jatropha Organization (IJO), Kuala Lumpur

IPS (2008) Jatropha key to self-sufficiency? Inter Press Service (IPS), Manila

Kapilkarn K, Peugtong A (2007) A comparison of costs of bio-diesel production from transesterification. Int Energy J 8:1–6

Kent A, Mercer D (2006) Australia's mandatory renewable energy target (MRET): an assessment. Energy Policy 34:1046–1062

Kim H, Kim S, Dale BE (2009) Biofuels, land use change, and greenhouse gas emissions: some unexplored variables. Environ Sci Technol 43:961–967

Kløverpris J, Wenzel H, Nielsen PH (2008) Life cycle inventory modelling of land use induced by crop consumption. Part 1: conceptual analysis and methodological proposal. Int J Life Cycle Assess 13:13–21

Latner K, O'Kray C, Jiang J (2006) Strict government control characterizes Chinese biofuel development. An online review of Foreign Agricultural Service initiatives and services. United States Department of Agriculture (USDA)

Lopez GP, Laan T (2008) Biofuels at what cost? Government support for biodiesel in Malaysia. One of a series of reports addressing subsidies for biofuels in selected developing countries. The Global Subsidies Initiative (GSI) of the International Institute for Sustainable Development (IISD), Geneva

Lund P (2009) Effects of energy policies on industry expansion in renewable energy. Renewable Energy 34:53–64

Magdoff F (2008) The political economy and ecology of biofuels. Monthly Rev July–August:34–50

Mathews J (2007) Biofuels: what a biopact between North and South could achieve. Energy Policy 35:3550–3570

MoE (2007) Policy DB Details: Thailand. Renewable energy and energy efficiency partnership. Ministry of Energy, Bangkok, Thailand

Molle F, Floch P, Promphaking B, and Blake DJH (2009). "Greening Isaan": Politics, Ideology, and Irrigation Development in Northeast Thailand. In: F. Molle, T. Foran, and M. Käkönen (eds). Contested Waterscapes in the Mekong Region: Hydropower, Livelihoods and Governance, pp. Earthscan, London, pp. 253–282

Molle F, Floch P (2008). Megaprojects and social and environmental changes: the case of Thai "water grid". Ambio 37:199–204

Naylor RL, Liska AJ, Burke MB, Falcon WP, Gaskell JC, Rozelle SD, Cassman KG (2007) The ripple effect: biofuels, food security, and the environment. Environment 49:32–43

Nguyen TLT, Gheewala SH (2008a) Life cycle assessment of fuel ethanol from cane molasses in Thailand. Int J Life Cycle Assess 13:301–311

Nguyen TLT, Gheewala SH (2008b) Life cycle assessment of fuel ethanol from cassava in Thailand. Int J Life Cycle Assess 13:147–154

Nguyen TLT, Gheewala SH, Garivait S (2007a) Energy balance and GHG-abatement cost of cassava utilization for fuel ethanol in Thailand. Energy Policy 35:4585–4596

Nguyen TLT, Gheewala SH, Garivait S (2007b) Fossil energy savings and GHG mitigation potentials of ethanol as a gasoline substitute in Thailand. Energy Policy 35:5195–5205

Nguyen TLT, Gheewala SH, Bonnet S (2008) Life cycle cost analysis of fuel ethanol produced from cassava in Thailand. Int J Life Cycle Assess 13:564–573

Nunt-Jaruwong S (2008) Thailand renewable energy policy – the potential for electricity from dendro-power (summary). Department of Alternative Energy Development and Efficiency, Bangkok

OECD (2008) Rising food prices: causes and consequences. Paper prepared for the DAC High Level Meeting, 20–21 May 2008

OEPP (2000) Thailand's Initial National Communication under the United Nations Framework Convention on Climate Change. Office of Environmental Policy and Planning, Ministry of Science, Technology and Environment (MOSTE)

Oxfam (2008) Another inconvenient truth: how biofuel policies are deepening poverty and accelerating climate change. Oxfam, London

Pleanjai, S., Gheewala, S. H. and Garivait S. (2004). Environmental Evaluation of Biodiesel Production from Palm Oil in a Life Cycle Perspective. in The Joint International Conference on "Sustainable Energy and Environment (SEE)

Pleanjai S, Gheewala SH, Garivait S (2009a) Greenhouse gas emissions from production and use of palm methyl ester in Thailand. International Journal of Global Warming (in press)

Pleanjai S, Gheewala SH, Garivait S (2009b) Greenhouse gas emissions from production and use of used cooking oil methyl ester as transport fuel in Thailand. Journal of Cleaner Production 17(9): 873–876

Peskett L, Slater R, Stevens C, Dufey A (2007) Biofuels, agriculture and poverty reduction. Natural Resource Perspectives 107. Overseas Development Institute (ODI)

Phalakornkulea C, Petiruksakula A, Puthavithi W (2009) Biodiesel production in a small community: Case study in Thailand. Resour Conserv Recycl 53(3):129–135

Praiwan Y (2008) Shell delays E20 on supply concerns: tight ethanol market may ease by year-end. Bangkok Post, Bangkok

Prakash A (2008) Cassava: international market profile. Background paper for the competitive commercial agriculture in Sub-Saharan Africa (CCAA) study. Trade and Markets Division: Food and Agriculture Organisation (FAO), Rome

Prasertsan S, Sajjakulnukit B (2006) Biomass and biogas energy in Thailand: potential, opportunity and barriers. Renewable Energy 31:599–610

Preechajarn S, Prasertsri P, Kunasirirat M (2008) Thailand Bio-Fuels Annual 2008 GAIN Report – Global Agriculture Information Network. Global Agriculture Information Network (GAIN), USDA Foreign Agricultural Service

Prueksakorn K, Gheewala SH (2008) Full chain energy analysis of biodiesel from *Jatropha curcas* L. in Thailand. Environ Sci Technol 42:3388–3393

Puppán D (2002) Environmental evaluation of biofuels. Periodica Polytechnica Ser Soc Man Sci 10:95–116

Rajagopal D, Zilberman D (2007) Review of environmental, economic and policy aspects of biofuels. World Bank, Washington DC

RCP (2006) The prototype of the community bio-diesel project. The Royal Chitralada Projects (RCP), The Crown Property Bureau (CPB), Thailand, Bangkok

Robertson GP, Dale VH, Doering OC, Hamburg SP, Melillo JM, Wander MM, Parton WJ, Adler PR, Barney JN, Cruse RM, Duke CS, Fearnside PM, Follett RF, Gibbs HK, Goldemberg J, Mladenoff DJ, Ojima D, Palmer MW, Sharpley A, Wallace L, Weathers KC, Wiens JA, Wilhelm WW (2008) Sustainable biofuels redux. Science 322(5898):49–50

Rossi A, Lambrou Y (2008) Gender and equity issues in liquid biofuels production minimizing the risks to maximize the opportunities. Food and Agriculture Organization (FAO), Rome

Rowe M (2008) Fuelling the debate. Geographical Dossier Special Report:44–51

Sarochawikasit R (2006) Renewable Energy Development and Environment in Thailand. DEDE, Ministry of Energy (MoE), Thailand, Bangkok
Scharlemann JPW, Laurance WF (2008) How green are biofuels? Science 319:43–44
Schmidt J (2008) System delimitation in agricultural consequential LCA. Int J Life Cycle Assess 13:350–364
SCOPE (2008) Biofuels: environmental consequences and interactions with changing land use. In: Proceedings of the scientific committee on problems of the environment (SCOPE) international biofuels project rapid assessment, Gummersbach, Germany, 22–25 Sep 2008
SEA-CR (2007) Field survey on sustainable palm oil in Thailand and compliance with international standards. Southeast Asia Consult & Resource Company Ltd, Chiang Mai
Searchinger T, Heimlich R, Houghton RA, Dong F, Elobeid A, Fabiosa J, Tokgoz S, Hayes D, Yu T-H (2008) Use of U.S. croplands for biofuels increases greenhouse gases through emissions from land-use change. Science 319:1238–1240
Sengmany P (2007) The environmental impacts of trade liberalization in the bio-diesel section of Lao PDR. National Council of Sciences, Lao PDR
Setiadi T, Husaini DA (1996) Palm oil mill effluent treatment by anaerobic baffled reactors: recycle effects and biokinetic parameters. Water Sci Technol 34(11):59–66
Sharlemann J, Laurance W (2008) How green are biofuels? Science 319:52–53
Shrestha RM, Pradhan S (2008) CO_2 Reduction in transport sector in Thailand: some insights. In: 3rd low carbon society workshop, Tokyo. Asian Institute of Technology, Bangkok
Shrestha RM, Pradhan S, Liyanage M (2008) Effects of carbon tax on greenhouse gas mitigation in Thailand. Climate Policy 8:140–155
Siriwardhana M, Opathella GKC, Jha MK (2008) Bio-diesel: initiatives, potential and prospects in Thailand: a review. Energy Policy 37(2):554–559
TDRI (1990) Enhancing private sector research and development in Thailand. Thailand Development Research Institute (TDRI), Bangkok
Tenenbaum DJ (2008) Food vs fuel: diversion of crops could cause more hunger. Environ Health Perspect 116:A254–A257
Uddin SN, Taplin R, Yu X (2006) Advancement of renewables in Bangladesh and Thailand: policy intervention and institutional settings. Natl Resour Forum 30:177–187
Uhlenbrook S (2008) Biofuel impacts on water: what do we know, and what do we need to know? UNESCO-IHE
United Nations Development Program (UNDP) (2007) Human Development Report. New York, USA
Varghese S (2007) Biofuels and global water challenges. Institute for Agriculture and Trade Policy, Minneapolis, MN
Woods Institute for the Environment (2006) The impacts of large-scale biofuel use on water resources. In: Workshop on the environmental, resource, and trade implications of biofuels, Woods Institute for the Environment, Stanford University, Palo Alto, CA
Wright T (2008) Firms back a plan to put the green in ‘green gold’. The Wall Street J 18 Jan 2008
WRM (2007) Brazil: Agro-fuels represent a new cycle of devastation of the Amazon and Cerrado regions. WRM’s bulletin N° 116, March 2007. World Rainforest Movement (WRM), Uruguay
WWAP (2009) United Nations World Water Development Report 3. World Water Assessment Programme (WWAP). Colombella (Perugia), Italy
Yoosin S, Sorapipatana C (2007) A study of ethanol production cost for gasoline substitution in Thailand and its competitiveness. Thammasat Int J Sci Technol 12:69–80
Zuurbier P (2008) Food and energy: impacts on land use. Biofuels and sustainability: Brazilian perspectives. Wageningen UR, The Netherlands

Chapter 7
Enabling Sustainable Shrimp Aquaculture: Narrowing the Gaps Between Science and Policy in Thailand

Dao Huy Giap, Po Garden, and Louis Lebel

Introduction

Since the early 1980s, shrimp aquaculture has expanded rapidly with production increasing more than 100-fold from 31,000 t in 1976 to 2.6 million tons in 2006 (Fig. 7.1). About 90% of farmed shrimp is produced in Asia (mainly in China, Thailand, Vietnam, Indonesia and India). The other 10% is produced mainly in Latin America, where Mexico, Brazil and Ecuador are the largest producers (Fig. 7.1). The two main aquaculture species that are farmed – black tiger shrimp (*Penaeus monodon*) and Pacific white shrimp (*Penaeus vannamei*) – accounted for about 88% of the shrimp in the aquaculture industry in 2006.

Today, almost half of the shrimp eaten is grown in tropical developing countries in coastal ponds and traded to be eaten in restaurants and homes in temperate industrialized countries. Thailand and Vietnam are the largest exporters, while much of China's growing production is consumed domestically. The Europe, US, and Japan are the largest importers, together comprising 80–91% of total global import volumes each year between 1976 and 2006 (Fig. 7.2).

The profitable expansion of the shrimp aquaculture industry can be attributed to the continuous improvement in aquatic production technologies combined with opportunities created by international trade which has created a global production–consumption system based on farms with high inputs, stocking densities and yields

D. H. Giap(✉)
Department of Agriculture, Hanoi University of Agriculture, Hanoi, Vietnam
e-mail: awuaasia@gmail.com

P. Garden
Internews' Earth Journalism Network, Chiang Mai, Thailand
e-mail: Po@internews.org

L. Lebel (✉)
Unit for Social and Environmental Research, Chiang Mai University, Chiang Mai, Thailand
e-mail: louis@sea-user.org

L. Lebel et al. (eds.), *Sustainable Production Consumption Systems: Knowledge, Engagement and Practice,* DOI 10.1007/978-90-481-3090-0_7,

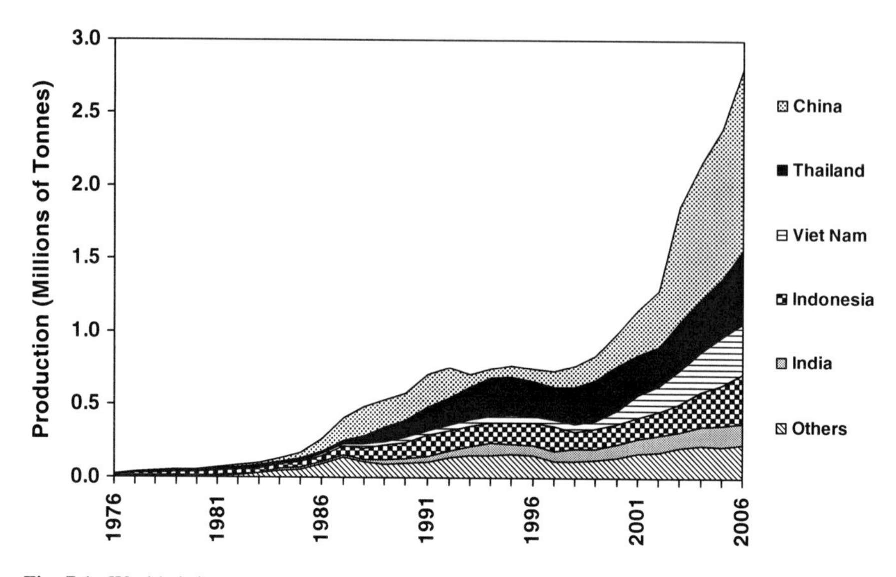

Fig. 7.1 World shrimp aquaculture production during 1980–2006 and world top five countries in 2006 (FAO 2008)

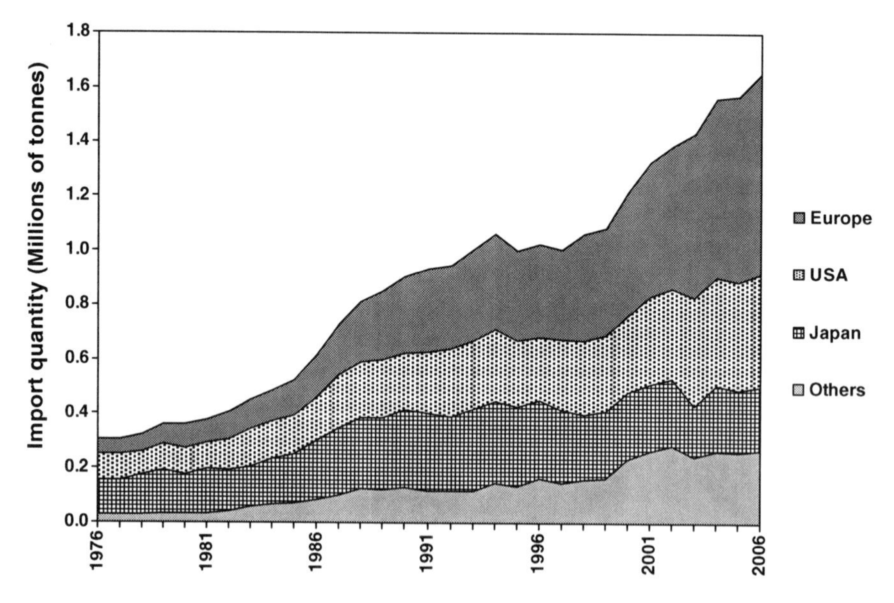

Fig. 7.2 Shrimp imports quantity by US, Europe, Japan and others during 1976–2006 (FAO 2008)

(Lebel et al. 2002). But it can also be attributed to subsidies from nature (Barbier et al. 2002; Naylor et al. 1998).

Shrimp aquaculture depends directly or indirectly on a range of coastal and marine ecosystem services (Deutsch et al. 2007; Naylor et al. 2000). It is assumed that wastewater and sediments disposed of in local waterways, creeks and coastal zone are assimilated by resilient local ecosystems. In some locations, rapid development of estuarine habitats for shrimp farming has clearly resulted in destruction and degradation of valued mangroves and other wetland habitats and affected access to natural resources by others (Bailey and Pomeroy 1996; Flaherty et al. 1999; Lebel et al. 2002; Neiland et al. 2001; Primavera 2006; Vandergeest et al. 1999).

To these environmental impacts can be added concerns about fairness in distribution of profits and risks within communities and along the commodity chain (Bailey 1988; Goss et al. 2000; Lebel et al. 2008; Vandergeest 2007). As a consequence of negative environmental and social impacts, the industry regularly faces public criticism and questions over its sustainability (Caffey et al. 2001; EJF 2003; Solidarity Center 2008; Stonich and Bailey 2000).

In response the shrimp aquaculture industry has repeatedly made commitments to improving pond management practices (FAO et al. 2006; NACA and FAO 2000. The Thai government has committed itself to higher standards and enforcement of regulations throughout the commodity chain. Certification schemes have been introduced to assure retailers and consumers, primarily that food is safe to eat, but increasingly that it is grown in environmentally and socially responsible ways (Lebel et al. 2008; Vandergeest 2007). At the same time innovations in rearing methods, the species cultured and post-harvest processing continue to emerge that require re-making links between science, policy and practices (Lebel et al. 2009a).

In this paper we focus on attempts to close gaps and address the tensions between science and policy in the pursuit of more sustainable shrimp aquaculture. We acknowledge that efforts to bring research-based knowledge to bear on the sustainability of the shrimp production–consumption system need be cognizant of at least three major goals driving behavior of key stakeholders: economic profitability, social development, and environmental health. We also accept that exactly what should constitute progress towards sustainability is frequently re-negotiated as new knowledge comes to light and powerful actors shift their perceived interests and expectations (Lebel et al. 2009a).

The paper is organized as follows. The next section on Research and Policy for Sustainability describes a selection of the sustainability issues associated with shrimp production–consumption system. We contrast what advice is being given to governments by researchers with the content and basis of policy. The section following this on Boundary Initiatives analyses several initiatives that can be construed as efforts to close the gaps between science and policy with respect to sustainability challenges in shrimp aquaculture. The final main section on Closing Science Policy Gaps synthesizes the findings from these earlier analyses.

Research and Policy for Sustainability

In this section we explore specific examples of gaps and tensions between policy and research in four issue areas: competitiveness, social benefits, environmental impacts, and resource use.

Competitiveness

As the shrimp aquaculture industry has expanded and more countries have brought significant areas into active production (Fig. 7.1) issues of competitiveness arise more and more frequently (Belton and Little 2008; Saengnoree et al. 2002; Sagheer et al. 2007). For many proponents in countries such as Thailand where the industry is long established, sustainability is equated with sustained profits. For other, newer producers, it is about establishment and expansion of market shares.

Government policy has been crucial to expansion of industry in Thailand (Brimble and Doner 2007; Goss et al. 2000; Ongsritrakul and Hubbard 1996; Sagheer et al. 2007). Apart from subsidies, adjustments to property rights, and provision of infrastructure in coastal areas, there has also been significant investment in research both through universities and quasi-public institutions like the Thai National Center for Genetic Engineering and Biotechnology (BIOTEC) (Tanticharoen et al. 2008).

Another relevant government agency has been the National Research Council of Thailand (NRCT). It published, for instance, a 3 year (2004–2007) shrimp aquaculture research strategy in the leading aquaculture business magazine "*Sart Nam*". The main objective of the strategy was to increase Thailand's competitiveness in the world shrimp market by improving productivity by 50% while reducing farming areas by one-half. A centralized research and development strategy was central to a plan that included a portfolio of projects worth about 700 million baht (US$17 million). This plan followed an earlier request for funds for biotechnology development to boost competitiveness in agriculture including shrimp farming (The Nation 2003).

In September 2008, a second roadmap was launched emphasizing a hygienic and green industry to add-value, maintain existing channels and gain access to new niche markets. Shrimp farmers, for example, were to be encouraged to replant mangroves and use seawater in ponds to absorb waste and residues (Keeratipipatpong 2008). Mr. Niwat Suteemechaikul, director-general of the Department of Fisheries argued that the "greening" would be very helpful for promotion of Thai shrimp in overseas markets.

One key factor in competitiveness has been the capacity of the Thai industry to speedily respond to concerns with food safety. In 2002, for instance, the EU discovered antibiotic residues in imported cultured shrimp from Thailand and banned shrimp imports. In response, the Thai Government stepped up shrimp quality inspection programs, including residue testing and implemented traceability measures. All farms are now required to adhere to Good Agricultural Practice guidelines and have to provide an Aquatic Animal Movement Document (Lebel et al. 2008; Phillips et al. 2003). The movement documentation starts from the hatchery stage and must be passed on to processors on sale of harvest. Exporters need to submit the documents to get export permissions. Some companies have resorted to using Radio Frequency Identification Devices to improve traceability in later stages by recording farm practices, feed uses and details of product transport and processing (Ekmaharaj 2006). This scheme was set up by the Department of Fisheries and the National Innovation Agency (NIV) under the Ministry of Science and Technology.

The NIV has also established an Intellectual Property Management Unit. The aim, according to NIA director Supachai Lorlowhakarn is to "work as an IP broker, matching local researchers pursuing valuable research and private companies that want to manufacture by-products from that research" (Rungfapaisarn 2006). Its first deal was with Kasetsart University working on photosynthetic bacteria that could raise productivity of shrimp farms. Organizations like BIOTEC use Thai and international patent systems to protect intellectual property (BIOTEC 2007).

But it has been the private sector which has been crucial. The Thai agri-business multinational Charoen Pokphand (CP) Group has been a key factor in Thailand's competitiveness in the shrimp industry. It played a major role in the early transfer and refinement of technologies from Taiwan and Japan. Pananond (2006) and others attribute much of the company's success to it's ability to combine technological capabilities in agri-business, established first in poultry, with networking skills. The CP Group's significant capacity for vertical integration in the shrimp aquaculture industry (Goss et al. 2000) has also enabled it to become a major regional shrimp exporter with bases not only in Thailand, but also Malaysia, Indonesia, Vietnam and India. CP could guarantee traceability of its products when many other firms could not (Pananond 2006).

The main contribution from technical research has been related to efficiency and risk. Most pond-level research has been aimed at improving the amount of shrimp produced per unit of inputs. Many informants, both farmers and within the bureaucracy in Thailand, told us how such research should draw much more on the experiences of farmers. Further down the chain, research and development to support value-adding in food processing has been another key factor in maintaining Thailand's competitiveness (Brimble and Doner 2007; Suwunnamek 1997).

Another area of critical research has been to reduce the risks of major disease outbreaks. This research is more multistage and multilevel. Ensuring that brood stocks are disease free has been a key objective from early on in the development of the industry. But only after the widespread availability and adoption of specific pathogen-free white shrimp (*Liptopenaeus vannemei*) more complete biosecure shrimp culture became feasible (Lightner 2005). At the farm level it comes down more to disease management protocols both to reduce risks of infection and also the spread of disease when it occurs (Menasveta 2002). Further down the chain food safety protocols, traceability measures and certification schemes are becoming key instruments (Lebel et al. 2008).

Research to help farmers better manage their investment risks is scarce. The competitiveness framing of sustainability in policy is at a "higher-level" which renders individual farmers largely invisible (Phillips et al. 2003). One example is a study of alternative water management systems that showed farmers would voluntarily switch to closed systems with lower short-term profits because of the longer-term benefits of more stable and predictable production (KASAI et al. 2005). Such research is needed as uncertainties about production due to weather and disease as well as costs of inputs or market prices make financial risk management at the farm level a critical skill. The needs seem particularly acute for smaller farms.

Social Benefits

In aggregate, national, terms the monetary value of the industry is huge. In Thailand, seafood exports contributed US$4.7 billion or 40% of total food export value of $12 billion. The total value of the shrimp chain within Thailand was $2.4 billion of which export value was $1.7 billion in 2004. Overall shrimp aquaculture contributed about 90% of total Thai shrimp production and export values. We have estimated elsewhere that shrimp aquaculture and its supporting industries created around 380,000 full-time jobs, and generated $1.4 billion value-added (Giap and Lebel 2006).

A standard commodity chain analysis (Giap and Lebel 2006) revealed, not surprisingly, that the benefits generated by shrimp aquaculture industry are very unequally distributed along the commodity chain. About 50% of total jobs created were shrimp growers but they receive only 11% of total value-added. By contrast, the 9.6% of jobs that were created in the U.S. shrimp trade industry shared 66.8% of the value-added (Fig. 7.3).

The structure and organization of the industry within Thailand reflects the power and leverage of different actors. There are about 30,000 farms. But most other stages are owned by a handful of major companies. For example, 95% of the shrimp feed consumed in Thailand was produced by ten companies. CP has more than 55% share of the shrimp feed market. There are about 130 processing plants in Thailand, and about 300 shrimp sellers operating in the auction market. With the shift from growing black to white shrimp even hatcheries have been "rationalized" to a much smaller number of players. Likewise there are only a few large buyers of "trash fish"

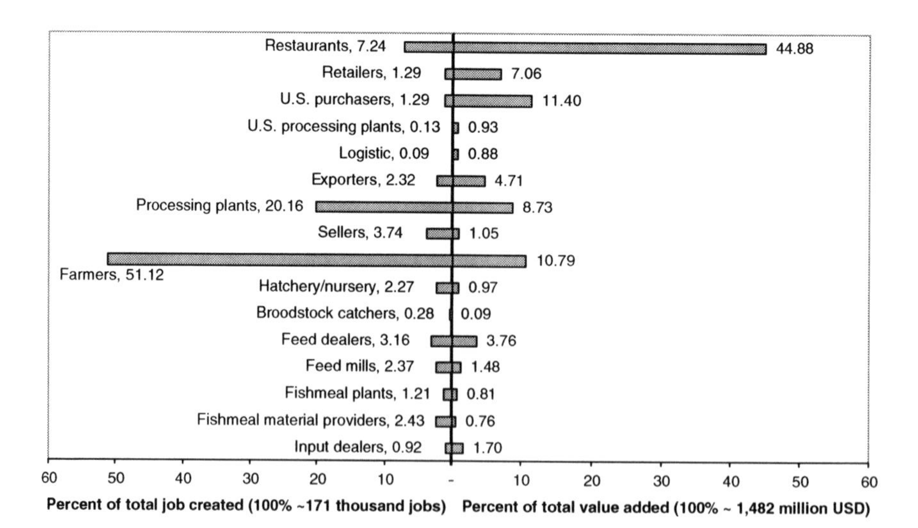

Fig. 7.3 Percent of total job created vs. percent of total value-added for different stages along the commodity chain from aquaculture-reared shrimp in Thailand through its trade and consumption in the United States

to make fishmeal so that apart from competition with sources outside Thailand they can strongly influence prices.[1]

The report "The True Cost of Shrimp" released by the non-governmental organization Solidarity Center (2008) documents cases of alleged child labour and abuses of migrant workers in the shrimp processing industry in Thailand. The report is consistent with an earlier study in 2006 by Institute for Population and Social Research of Mahidol University and supported by the International Labor Organization (ILO). Thailand's Ministry of Commerce strongly denied the allegations and defended the shrimp and seafood industries' adherence to international standards (QFFI 2008). The Global Aquaculture Alliance was also quick to counter questions in the report about the viability of its certification schemes (QFFI 2008).

Shrimp farming earns foreign income and creates some jobs. But does shrimp aquaculture contribute more to social development in coastal areas than alternatives like tourism, sustainable use of mangrove and wetland ecosystems, or other agricultural activities? Researchers, beginning with Bailey (1988) have pointed out some of the problematic features with respect to property rights and access to mangrove and coastal wetlands in the areas where intensified shrimp production methods were first introduced (Bailey and Skladany 1991). He also has drawn attention to close connections between the state and private sector and multilateral agencies (Bailey 1988) commenting on the bias against other coastal land uses such as agricultural production and coastal wetlands that would better serve the livelihoods of the poor (Bailey and Pomeroy 1996; Stonich and Bailey 2000).

Aquaculture can take away access to coastal resources from poor families who depend on them. Poor households in a coastal mud-flats area in Vietnam, for example, lost access to land taken for the farming of shrimp and clams (Dao et al. 2007). In southern Thailand we also identified highly vulnerable households for whom the loss of access to coastal areas including in coastal creeks and mangroves for collecting marine products and fishing affected their livelihood security. Some of the most significant ecological impacts for local inhabitants were sedimentation and pollution of coastal and mangrove creeks and the salinization of farming areas and wells used for potable water (Lebel et al. 2008). At the same time, people find ways of coping: we observed older women collecting 'escaped' and 'discarded' shrimp in drainage canals for sale in local markets.

Research and policy have not done all that much to understand and address the concerns of disadvantaged groups not directly involved in shrimp aquaculture in local communities.

Environmental Impacts

Among several issues, waste water and sediments that are improperly treated and disposed of are one of the direct environmental impacts of shrimp farming operations.

---

[1] Interview, Mr. Amnuay Oerareemitr, Deputy President of the Thai Fishmeal Producers Association.

The effluents of shrimp farm generally contain high levels of Biological Oxygen Demand (BOD), nitrate, phosphate, chlorophyll-a, and bacterial concentrations and may also carry residual chemicals and drugs used in pond cleaning and disease control (Graslund et al. 2003; Holmstrom et al. 2003). Environmental health risks, including increased resistance to antibiotics and exposure of farm workers and local residents to chemicals used in aquaculture are under-studied and thus not adequately understood (Sapkota et al. 2008).

Salinization of land and water (Braaten and Flaherty 2001) can reduce rice yields (Kiatpathomchai et al. 2008) and can become a source of conflict with other water and land-users (Flaherty et al. 1999). Untreated effluent discharges by shrimp farms are one of the principal causes of disease epidemics which may also effect natural populations (Neiland et al. 2001). Sediments dumped into waterways and coastal creeks affect the growth and survival of mangroves (Vaiphasa et al. 2007). Poor water quality also reduces shrimp production; shrimp farms can also be adversely affected by pollution arising from other agricultural and urban or industrial land- and water-uses (Neiland et al. 2001).

Expert advice to growers has been to limit water exchanges and improve the quality of effluents (Boyd et al. 2005, 2007). Disease prevention can benefit from greatly reduced water exchange (Menasveta 2002). There is also a large body of formal research-based and more practice-oriented knowledge about improving farming practices so that less feed is wasted and effluent problems are reduced at their source. Some of these are discussed in the section below on resource use.

At the policy level governments have introduced standards aimed at securing better water and pond sediment management practices. Policies included requiring treatment of waste water through settlement basins or other methods (Tacon and Forster 2003). An on-going, but still not well enough understood issue, is why better management practices are yet to be more widely adopted in the field. The reasons are partly economic incentives (Stanley 2000), especially in the case of smaller farms, but also related to capacity and knowledge of farmers. More research on the effectiveness of different programs to bring about changes in practices is needed as optimism and pessimism with respect to changes in the "real world" is one of the key divides among coalitions either supporting or skeptical (Bene 2005) of the future sustainability of the shrimp aquaculture industry.

Resource Use

Concerns about environmental impacts should not be restricted to ponds and their immediate surrounds (Lebel et al. 2002). Inputs from marine ecosystems are the foundations of shrimp feed (Deutsch et al. 2007; Naylor et al. 1998, 2000). Our analysis of material flow of fishmeal in Thailand, for example, shows that about 4 kg of fish is required to produce one kg of fishmeal (Giap and Lebel 2006). Fish used for fishmeal production were "trash fish", fisheries by-products and the so-called low-value fish varieties. The expansion of aquaculture and processing plants has increased the amount of fisheries by-products in the ingredients for the fishmeal industry. We note for example that about 65% of raw shrimp is edible; the other 35% is waste or by-products and can be-reused.

Thailand's level of fishmeal import dependency rose and the number of import sources increased between 1980 and 2000 (Deutsch et al. 2007). Thailand utilizes fishmeal sources all over the globe, importing from the established major exporters such as Denmark and Chile. Ties are also maintained to regional sources throughout the period in both the number of sources (7 of 14) and volumes (22%) of total imports. In 2000, for instance, Thailand was highly dependent on one major supplier, Peru (72% of imports).

Shrimp require high protein content fishmeal. This can be imported or produced from relatively higher quality fish (typically classified as low value food fish). In Thailand, the revenues from trash fish and low value fish only contribute about 5% of the value of the total catch. Mr. Amnuay Oerareemitr, Deputy President of the Thai Fishmeal Producers Association told us that higher prices for trash fish for fishmeal encourage fishermen to catch and store on ice to preserve freshness and sell to fishmeal mills rather than discard at sea or use as fertilizer. The consequences of by-catch use are complex but potentially significant, as up to 27% of total catch is discarded. While we acknowledge that the environmental issues associated with by-catch are probably best addressed through fisheries management, it is worth noting the impacts of demand from the animal feeds industries (not only for shrimp). There is a need to trace the marine resource base of aquaculture products and make sure we can detect feedbacks even at distant and short duration to avoid over-exploitation (Deutsch et al. 2007).

A more distant but related issue is the additional competition that harvesting for fishmeal places on poorer coastal fishers who may depend directly on some of these species to meet protein needs in their diet. Reducing the amount of fishmeal used by shrimp industry could help reduce pressure on natural stock, but proper fisheries management and economic incentive to distribute fish protein to poor people are also key issues.

Intensive aquaculture systems require inputs, but so do alternatives such as harvesting wild shrimp. We estimated that catching wild shrimp requires about twice the amount of energy as aquaculture-produced shrimp. About 50% of revenues from wild-caught shrimp go to fuel costs. On average, 1 kg of wild-caught shrimp requires 2 L of diesel fuel to catch, whereas 1 kg of aquaculture-reared shrimp requires the equivalent of 1 L of diesel fuel mostly for feed and pond maintenance (Giap and Lebel 2006). In 2004, 66% of Thai shrimp production was from *P. vannamei*, 26% from *P. monodon* and the balance 8% comprised *Macrobrachium rosenbergii* and other shrimp species. Previously most shrimp production in Thailand consisted of *P. monodon* gravid females which had to be harvested from the wild. The harvesting required about 400 L of diesel fuel per 1 kg of shrimp brood stock underlining that energy needs may be significant beyond the outgrowth pond stage (Giap and Lebel 2006).

Several studies have now attempted to assess the footprints of shrimp farms and the full life cycle impacts of producing shrimp (Mungkung 2005; Saowalak 1999a, b).

The expert advice has ranged from suggestions to become more efficient to abandoning consumption of shrimp entirely. On the feed side, leading experts often point to the potential and limitations for plant-based proteins to substitute for those derived from ocean-caught organisms (Tacon and Forster 2003). Turning these analysis into practical advice for policy, however, requires additional work as not all important impacts can be quantified and thus used to set standards or labelling criteria (Mungkung 2005).

Some researchers have suggested using indicators to evaluate performance of aquaculture (Boyd et al. 2007). Those covering resource uses might include measures of efficiency with which feed, protein, nutrients, fishmeal, water, land and energy are used. Others have calculated use of ecosystem services in different ways, for example, estimating the amount of mangrove area needed to absorb waste nutrients (Shimoda et al. 2007).

Although issues of resource use and more distant ecosystem impacts have not been central to policy to make shrimp aquaculture sustainable, this appears to be changing with an increasing interest in certification schemes. Consumers and those who retail shrimp to them anticipate marketing benefits from, for instance, carbon footprints.

Boundary Initiatives

The advice being given to Thai governments by researchers varies in form and diversity across issues. The rationale and content of respective government policies is also mixed reflecting how sustainability is understood and negotiated.

Thai researchers looking for funds from Thai sources, for instance, are usually encouraged to give attention to improving competitiveness of Thai firms. Foreign researchers and Thai researchers funded from other sources are open to a greater diversity of framings and approaches to sustainability, including social dimensions related to labor and local communities. Water pollution is largely understood as a technical and regulatory matter; the gaps are in getting incentives, information and monitoring right in the field. Resource use issues, in particular those related to fishmeal have hardly been addressed within Thailand, except in fringe experiments with alternative, locally produced feeds.

Tensions and gaps persist within and between science and policy in Thailand. Initiatives aimed at addressing these have arisen several times in different nodes and alliances in the production–consumption system. In this section we look at five boundary initiatives and their impacts on Thai policy-making and research.

Private Industry Associations

The Thailand shrimp industry is comparatively well connected by trade associations organized along the supply chain, from fishmeal and hatcheries, through farmers to frozen food processors and exporters (Suwunnamek 1997). These associations are important to addressing science-policy gaps in several ways with varied implications for both knowledge and power.

Firstly, the associations are important for policy influence as entities that can represent and lobby the government on behalf of their constituencies. The large firms involved in food processing are particularly well-connected to government and are thus able to carefully look after their trade-related interests. The power of the so-called "Hong Yen" group to influence market conditions and policy is a frequent source of concern and complaints by farmers. A successful leader and farmer told us:

> Frozen food processors are probably the biggest enemy to the sustainability of shrimp aquaculture. All they do is cut costs and that is our money. Frozen food processors and exporters could not tell us in advance what sizes get the best prices and when there are problems such as EU regulation or antidumping laws- they always exaggerate it to cut prices.

Another related a similar story:

> If the industry is going to survive – something must be done about Hong Yen. They are the worst. They say they want small shrimp and once you get there they say they want big shrimp. You call and they give one price and when you get there they give another price. They could find so many reasons to cut the prices. If it is not the standards - if it is not COC then it is GAP. Things should be made clear and Hong Yen needs to be regulated. They play up news about anti dumping and they play up news about antibiotics and at the end nothing really happens. They dictate the prices at their will.

Secondly, associations as knowledge arenas help in gaining understanding where practical experience, innovations, and specific expertise are brought together and validated. Shrimp growers' associations are particularly good example, with their seminar culture and supporting networks of smaller groups or "Chum Rom" (Lebel et al. 2008). The Surat Thani Shrimp Growers' Association established in 1990, for example, initially focused on reducing costs of inputs, but later became active in promoting conservation and social responsibility (CORIN 2000). Local associations are also important to shrimp farmers as they are at the intersection of several networks and provide a place where timely information can be shared and potential mis-information from suppliers and buyers can be edited (Lebel et al. 2008).

A national-level Thai Marine Shrimp Farmers' Association (TSA) was established in 1996 and deals with national and international issues, for example, being a member of the Global Aquaculture Alliance (see section on International Certification). Surasak Dilokiet, Secretary of TSA told us that "The research direction of our group focuses on shrimp farmers and not shrimp. Their main question is what should shrimp farmers do in 3–5 years time given their expertise and long experience with aquaculture. That is what we really want to know".

Thirdly, the associations work at stimulating innovation by acting as future consumers of new products, technologies and services. Associations provide a way for firms to invest in and sponsor innovative research that is likely to be of system-wide benefit. Of course individual firms retain other knowledge for themselves which is crucial to their individual competitive advantages. Thus, one use that farmers have for researchers is to help describe and communicate their experiences to others. Pinyo Kiatpinyo, President of White Shrimp Grower Association said:

> Science is a crucial tool in learning how to grow shrimp successfully. The key is to network and learn from each other. Use a professor to lead and to facilitate the process. The challenge is in how we could best use scientists – pick their brains and have them work for us. I suggest that a national institute might not work well but break it down and use a group of academics who are located in your region. If you live in Nakorn Nayok, get a group together and work with Kasetsart University. If you live in Chonburi, use somebody in Burapha University.

But other farmers are more skeptical of the technical emphasis of some researchers. Daycha Bunloedej a well-known black tiger prawn farmer from Prachuap Khiri Khan told us:

> Do not trust academics or the science they produce but use them to compliment one's own experience.... Perhaps scientists are not solving our issues that need their attention. Shrimp in nature can live with the virus: the question is how and why. What in nature gives it immunity? Nature will always have something to do with sustainable shrimp aquaculture.

Challenges like this from users of knowledge are of course exactly what are needed in pursuits of sustainability; associations provide settings or arenas where such exchanges can regularly take place.

Some firms like CP are so large and directly engaged with farmers that they behave like their own associations. CP has used demonstration farms and practical workshops to recruit and support growers who then buy their feeds – the company's main business. These contract farming relationships build capacity and help discipline farmers into keeping records. The contract group or clubs are another important feature of the innovative Thai shrimp industry. Thus when major new innovation emerges, like the switch from black to white shrimp (Lebel et al. 2009b), those in contract clubs with CP were at a distinct advantage in gaining access to post-larvae initially in very limited supply (Rath 2008; Turakit Tanh Setakit 2004).

Quasi-Public Agencies

The Thai National Center for Genetic Engineering and Biotechnology (BIOTEC) has been a significant actor in the knowledge system around shrimp aquaculture within Thailand. First established in 1983 it became one of the Centres of the National Science and Technology Development Agency in 1991. In 2006 it had almost 400 researchers and an annual budget of US$27 million (BIOTEC 2007).

BIOTEC has facilitated several initiatives related to shrimp. One good example is CENTEX Shrimp[2] set up by Mahidol University with support from BIOTEC and building on an earlier private-public consortium that had developed DNA probes

[2]For more on CENTEX see: http://www.biotec.or.th/centex/centex.htm and http://www.mahidol.ac.th/mueng/centex_shrimp.htm

for shrimp viruses (Brimble and Doner 2007). The CP Group, for example, donated 5 million baht to their activities in 2003 (Charoenwong 2003). CENTEX has maintained close relationships with business.

In 2000 BIOTEC set up the Shrimp Biotechnology Business Unit to commercialize findings from research, like the virus test kits in "Ezee Gene" series. Revenues are reinvested in research.

One of the key research areas have been in raising specific pathogen-free tiger shrimp. This has been a joint activity between the Shrimp Culture Research and Development Co. Ltd and Mahidol University (BIOTEC 2007). BIOTEC has also supported the establishment of the Shrimp Genetic Improvement Center in collaboration with Mahidol and Prince of Songkla Universities and the Royal Thai Navy in southern Thailand (BIOTEC 2007; Xinhua News Agency 2004). BIOTEC has provided funding for, and supported cooperation among, different parts of the commodity chain (Tanticharoen et al. 2008). This includes a network of farms and hatcheries with which there is regular collaboration. The largest role has been, in terms of technical areas, with respect to diagnostic tools for detection of viruses, genetic research and domesticated shrimp stock.

Code of Conduct

The Department of Fisheries (DOF) of Thailand has played a leading role in the development of Code of Conduct (CoC) for shrimp aquaculture farms. Successive versions of the code have become an arena in which research and policy can interact.

The Code of Conduct (CoC) for responsible shrimp aquaculture launched in Thailand in 1998 was based on FAO's Code of Conduct for responsible fisheries, ISO14001 and HACCP. Siri Tookwinas, a marine shrimp specialist, a key person in developing CoC in Thailand for farms and hatcheries, told us in jest that the impetus came from the EU white paper: "it seemed that they are concerned that our shrimp farmers are not well dressed".

Various associations in the industry backed the formulation of the CoC which was to be a sort of premium grade certification complimenting Good Agricultural Practices (GAP) which later was adopted as a minimum, food safety-oriented, standard. A history of long and close collaboration on science and technology issues in the shrimp industry between DOF and Kasetsart University provided important background knowledge into adapting the code. In the preparation of the CoC, shrimp clubs and associations in various provinces were consulted. Those inputs led to modest changes in some of the details of the code as the DOF's aim was "to create as little management practice change as possible for farmers to meet the standard".[3]

[3]Interview, Siri Ekmaharaj (Tookwinas), then with Department of Fisheries.

For many years the ability to bridge gaps between policy and science were not matched by the same capacity to take knowledge into practice. In surveys carried out in late 2000 in southern Thailand (Lebel et al. 2002), only half of the total number of shrimp farmers seemed to have heard of the main elements of the CoC. There were many initial difficulties in implementation of the CoC scheme. One of the main problems is that small farms do not have land or resources for water treatment facilities.[4]

In the last few years several alternative sets of standards and certification schemes have been proposed and promoted (Lebel et al. 2008). The Department of Fisheries has been defensive and dismissive of most of these endeavors. Representatives in meetings usually argue that their procedures are the most appropriate for the Thai context and alternatives are either impractical or redundant.

Apart from technical reasons other interests are also involved. The Department of Fisheries has a strong interest in protecting Thailand's competitiveness in shrimp aquaculture. One way it sees its role is in keeping some control and authority over standards and certification schemes.

Networks and Guides

The Network of Aquaculture Centres in Asia-Pacific (NACA) as an intergovernmental organization was established in 1988. It had an earlier existence as a UNDP project following support for its creation at FAO meetings since 1975 (UNDP & FAO 1992). The Secretariat of NACA is now hosted in the Department of Fisheries, Kasetsart University in Bangkok. The network includes research centres in 17 countries.

NACA joined a consortium on shrimp farming and environment in 1999. Initial partners were the World Bank, FAO and World Wildlife Fund and later joined by the United Nations Environment Program (UNEP). The group supported case studies and consultations leading to a major initial synthesis in 2002 (World Bank et al. 2002). Although draft copies were made available on the web for public comment through the NACA website, few of the major critics of the shrimp industry took up the task (Bene 2005). The consortium continued its work and produced a major set of guidelines for the shrimp industry in 2006: "The International Principles for Responsible Shrimp Farming" (FAO et al. 2006). The principles are intended as a guide and foundation for planning and management by public and private sectors (Table 7.1). NACA staff played a major role in preparing the various reports.

Work under the consortium program on Shrimp Farming and the Environment continues at NACA. Recent guides, handbooks and spreadsheet tools cover things like health and disease management, diet formulation and hatchery management practices. NACA also had some significant collaboration with aquaculture researchers

---

[4]Interview, Siri Ekmaharaj (Tookwinas), then with Department of Fisheries.

Table 7.1 International principles for responsible shrimp farming (FAO et al. 2006)

Principle	Explanation
Farm siting	Locate shrimp farms according to national planning and legal frameworks in environmentally suitable locations, making efficient use of land and water resources and in ways that conserve biodiversity, ecologically sensitive habitats and ecosystem functions, recognizing other land uses, and that other people and species depend upon these same ecosystems
Farm design	Design and construct shrimp farms in ways that minimize environmental damage
Water use	Minimize the impact of water use for shrimp farming on water resources
Brood stock and post-larvae	Where possible use domesticated selected stocks of disease free and/or resistant shrimp brood stock and post-larvae to enhance biosecurity, reduce disease incidence and increase production, whilst reducing the demand for wild stocks
Feed management	Utilize feeds and feed management practices that make efficient use of available feed resources, promote efficient shrimp growth, and minimize production and discharge of wastes
Health management	Health management plans should be adopted that aim to reduce stress, minimize the risks of disease affecting both the cultured and wild stocks, and increase food safety
Food safety	Ensure food safety and the quality of shrimp products, whilst reducing the risks to ecosystems and human health from chemical use
Social responsibility	Develop and operate farms in a socially responsible manner that benefits the farm, the local communities and the country, and that contributes effectively to rural development, and particularly poverty alleviation in coastal areas, without compromising the environment

at the Asian Institute of Technology, including a recent project to help small-scale shrimp farmers meet EU standards.

The impacts of the group's work are very significant regionally; within Thailand the impacts more narrowly on policy and innovation in the shrimp industry are less easy to trace, but still likely to be important.

Several NACA researchers, in publications, presentations and interviews continue to express concerns about and interest in helping small-scale farmers succeed in the shrimp industry. This is an intriguing concession and compliment to the consortium's emphasis on best management practices which often pose significant challenges to small farmers. Dr. Mohan, a NACA Aquatic Animal Health Specialist told us NACA's roles were to develop principles and better management practices (BMPs) for shrimp farming that build awareness and capacity of small-scale farmers to produce shrimp in a responsible manner and meet new standards and import requirements. NACA works closely with producing countries to develop BMPs, and according to them the number of farmers adopting BMPs has increased significantly since 2002 in many of the NACA member countries (e.g. India, Vietnam, Thailand and Indonesia).

International Certification

The Global Aquaculture Alliance (GAA) established in 1997 is an international, nonprofit trade association dedicated to advancing environmentally and socially responsible aquaculture (Lee and Connelly 2006). It has its headquarters in St. Louis, MI.[5] According to Dr. Lee, a GAA's Best Aquaculture Practices Standards Coordinator, it was formed, in part to draw criticism of NGOs away from buyers to the association where they could be more effectively handled (D. Lee and J. Connelly, 2006, personal communication).

The Aquaculture Certification Council (ACC), Inc. was established in 2003 with the help of GAA, as an independent, nongovernmental body to certify social, environmental and food safety standards at aquaculture facilities throughout the world (Lee and Connelly 2006). The ACC applies elements of the GAA Responsible Aquaculture Program through a certification system that combines site inspections and effluent sampling with sanitary controls, therapeutic controls and traceability. The GAA Responsible Aquaculture Program, mentioned above was derived from the FAO Code of Conduct for Responsible Fisheries.

The knowledge foundations of the code of practice developed by GAA included commissioned work in 1997 from well-known experts like Claude Boyd (Auburn University) and Michael Phillips (NACA). Claude Boyd was also a consultant for the Thai CoC (Ekmaharaj 2006). He emphasised in an interview that good management procedures can protect the environment from negative impacts of aquaculture, and that good practices are usually more profitable. He believes that aquaculture certification will ultimately lead to more efficient and responsible production practices (Claude Boyd, 2006, personal communication).

Part of ACC's mission is to help educate the aquaculture public regarding the benefits of applying best management practices and the advancing scientific technology that directs them. By implementing such standards, program participants can better meet the demands of the growing global market for safe, wholesome seafood produced in an environmentally and socially responsible manner. This includes countering claims made in reports by non-governmental organizations, for example, by pointing to lack of evidence, or out-of-date data, and improvements in practices (QFFI 2003, 2008).

The activities of GAA and ACC are supported by large shrimp buyers in the US like Darden Restaurants, operators of Red Lobster chain and Wal-Mart. Wal-Mart, the world's biggest retailer, requires those who sell shrimp to them to be certified by ACC. GAA helped forge the deal with Wal-Mart. The circularity of relationships is of obvious concern and was quickly noted by the Mangrove Action Program (MAP 2006). In Thailand, there has been some resistance and concern that the ACC's activities could effect the competitiveness of Thai shrimp (Hudson and Watcharasakwet 2007; The Nation 2007). Shrimp growers resent being

[5]For more on GAA see: http://www.gaalliance.org/

portrayed by western media as primitive or lacking in capacity (Hudson and Watcharasakwet 2007). The Thai government would rather have more influence over the certification process.

Jim Enright, a coordinator for MAP based in Thailand, told us in an interview in May 2006 that "certification needs to take into account the concerns of local communities; they should have inputs into standards, procedures and monitoring". Vandergeest (2007) points out that the ACC standards have provisions for local involvement including the need to comply with local laws, but not for local input into the technical standards.

The Global Aquaculture Alliance is also pro-active in defending its certification schemes, most recently for Best Aquaculture Practices in seafood processing, countering an NGO report about child and migrant labor in Thailand (QFFI 2008).

Despite resistance from Thailand's Department of Fisheries to the alternative ACC certification scheme, between July 2007 and September 2008, 13 farms in Thailand had sought and achieved ACC certification. With alternative arenas available, farmers who can are making choices of what they believe will best fit their business interests, for example, in gaining or securing access to particular market channels.

Closing Science Policy Gaps

Efforts to close gaps and reduce tensions between science and policy have taken different approaches and with mixed success.

Codes, standards and guidelines have been most visible outcome of efforts to get scientific understanding in a form relevant to practices and specifiable in policy. Much of the effort behind the initiatives taken by the Thai Department of Fisheries, NACA and GAA has been in synthesizing research understanding in a way that is palatable and practical for the industry. While each initiative has some of its own emphasis, for example, to the extent to which industry, government or researchers lead, all are consistent with a managerial, best practices, discourse (Bene 2005).

Consultation, dialogue and partnership have been an important feature of more successful parts of each initiatives, but rare overall. Consultations with industry were a feature of the development of the Code of Conduct, but dominated by downstream processing representatives. The GAA initiative has been affected by similar constraints. The consortium which NACA was intimately involved in has also tried to be inclusive and transparent, but its very framing of sustainability issues still tends to leave some groups of stakeholders out. World Wildlife Fund, recently held a dialogue on "their" certification scheme which emphasizes environmental issues, but struggled to involve farmers other than a few from the largest firms and associations. Going beyond the pond is crucial to initiatives aimed at enabling more sustainable PCS but ignoring the pond entirely is not an option either.

Key individuals and lead agencies help in bringing agendas forward. The first three initiatives discussed in section on Boundary Initiatives each start from the premise that sustainable shrimp aquaculture is possible and that better pond management practices are key. One reason for the similarities is that key experts, sometimes acting as consultants, were often the same. Although our analysis in this chapter has emphasized institutional aspects the role of individuals cannot be ignored. Another reason is that the various guides, standards and certification schemes although allegedly having different purposes, are in fact in competition with each other for the "intellectual space" and keep borrowing strong elements from each other. The lead agencies behind them have an "interest" in seeing their scheme adopted and not left behind.

At least four kinds of challenges to gap closure in the shrimp aquaculture remain important in the Thai context. Concisely these might be called the challenges of meaning, scale, uncertainty and justice.

First, a polarized, ideological, divide between those who believe progress towards sustainability is possible with better practices and others who think more fundamental changes to political economy are required (Bene 2005; Vandergeest et al. 1999). This gap is not so much between science and policy as between mixed coalitions of each. Within Thailand most research and policy has defended the industry, but with some differences in perspective on techniques or on whether and how to support smaller versus larger-scale farmers.

Second, a disjunct between understanding ways to improve farming practices and how they might be guided and regulated on the one hand, and relating this understanding to issues in other parts of the commodity chain, such as processing and exports, in a way that innovations strengthen not weaken competitiveness. Shrimp growers are usually portrayed as either the culprits or victims of unsustainability in the shrimp PCS when the behavior, influence and knowledge of other actors are much less scrutinized. This gap occurs both in thinking about, and acting on, farm- and sector-level or national-level issues and interests.

Third, an appreciation of the uncertainties in production and profits at farm-level and boom-bust cycles at national-level have not been translated into effective risk management policies or strategies. Instead there are knee-jerk policy responses to latest disease epidemic or fall in prices due to oversupply. This gap arises, in part, because the issue falls neatly in the crack between private and public interest and, in part, failures to take an adequately long-term perspective in both research and policy.

Fourth, continuing neglect of social issues by both research and policy. Mainstream agendas do not emphasize how risks, burdens and benefits are allocated along the chain and among places. Thai policy and research has largely reframed sustainability issues, as has been observed more globally (Bene 2005), as being largely technical problems of better pond management, devoid of more troublesome questions of interests, power and justice in the organization of the commodity chain or management of coastal zone development. The social development benefits of shrimp aquaculture are not adequately evaluated or compared.

The sustainability issues and science-policy initiatives considered in this chapter also illustrate the diversity of limitations faced by different kinds of actors. International epistemic networks may be very good at generating relevant knowledge but that is hard to channel into policy and practice unless their framing of issues is already part of the domestic policy and research agendas.

Governments usually look to centralization as a way to achieve sought after integration and then fail to benefit from incentives in more competitive and place-tailored contexts. On the other hand experiments with private-public partnerships and hybrid organizations have, on technical issues, often done well. Large firms and industry associations are good at building coalitions important for lobbying for investments in research and other support from government, but find it hard to turn-around and negotiate in more mixed-interest forums that include a much wider diversity of stakeholders and social development issues.

Conclusion

Government policy has contributed significantly to successful integration of science and industry in the shrimp aquaculture production–consumption system within Thailand. Most of the key technological and business management innovations that might be said to enable sustainability have been side-effects of changes driven by narrower concerns about shrinking profit margins and competitiveness.

In contrast, the social dimensions of the industry's impact remain relatively poorly understood by research and inadequately addressed by policy. Aggregate export statistics are insufficient grounds for considering issues of allocation of benefits and burdens among different social groups including those without an immediate direct stake in the industry. The neglect is not because these issues are unimportant, but because of a framing of performance and sustainability that excludes their consideration. Better pond management practices, better coastal zone land-use planning, and technical innovations in feed and improvements to food safety are expected to deliver the needed progress towards sustainability.

Significant gaps and tensions between science and policy remain in most areas. Going further will require a different style of policy-making and research agenda setting than has been typical in the past. Consultation, dialogue and partnership were the keys to initiatives that achieved at least some modest success. Much more scope for exploring alternatives and deliberation of priorities is needed, and these should involve not just bureaucrats and experts, but also shrimp farmers and other members of local communities.

Acknowledgements The work in this chapter was supported by grants from the David and Lucille Packard Foundation, the U.S. National Oceanic and Atmospheric Administration's Office of Global Programs.

References

Bailey C (1988) The social consequences of tropical shrimp mariculture development. Ocean Shoreline Manag 11:31–44

Bailey C, Pomeroy C (1996) Resource dependency and development options in coastal southeast Asia. Soc Natl Resour 9:191–199

Bailey C, Skladany M (1991) Aquacultural development in tropical Asia: re-evaluation. Natl Resour Forum 15:66–73

Barbier EB, Strand I, Sathirathai S (2002) Do open access conditions affect the valuation of an externality? Estimating the welfare effects of mangrove-fishery linkages in Thailand. Environ Resour Econ 21:343–367

Belton B, Little D (2008) The development of aquaculture in Central Thailand: domestic demand versus export-led production. J Agrarian Change 8:123–143

Bene C (2005) The good, the bad and the ugly: discourse, policy controversies and the role of science in the politics of shrimp farming development. Dev Policy Rev 23:585–614

BIOTEC (2007) Annual Report 2005–2006. The National Center for Genetic Engineering and Biotechnology (BIOTEC), Bangkok

Boyd CE, McNevin AA, Clay J, Johnson HM (2005) Certification issues for some common aquaculture species. Rev Fisher Sci 13:231–279

Boyd CE, Tucker C, Mcnevin A, Bostick K, Clay J (2007) Indicators of resource use efficiency and environmental performance in fish and crustacean aquaculture. Rev Fisher Sci 15:327–360

Braaten R, Flaherty M (2001) Salt balances of inland shrimp ponds in Thailand: implications for land and water salinization. Environ Conserv 28:357–367

Brimble P, Doner R (2007) University-industry linkages and economic development: the case of Thailand. World Dev 35:1021–1036

Caffey RH, Kazmierczak RF, Avault JW (2001) Developing consensus indicators of sustainability for Southeastern United States Aquaculture. Louisiana State University Agricultural Center, Baton Rouge, LA

Charoenwong P (2003) Improving parent breeders of the black tiger prawn: CPF helps support new BIOTEC Center. Charoen Pokphand Foods Plc. (CPF) Monthly Newslett 3:1–2

CORIN (2000) Shrimp farming experiences in Thailand: a continued pathway for sustainable coastal aquaculture. Coastal Resources Institute, Prince of Songkla University, Hat Yai, Thailand

Dao PTA, Hong PN, Tuan LX, Tho NH (2007) Xuan Thuy National Park: socio-economic status. MERC-MCD, Hanoi, Vietnam

Deutsch L, Gräslund S, Folke C, Troell M, Huitric M, Kautsky N, Lebel L (2007) Feeding aquaculture growth through globalization: exploitation of marine ecosystems for fishmeal. Global Environ Change 17:238–249

EJF (2003) Smash and grab. Environmental Justice Foundation, London

Ekmaharaj S (2006) Aquaculture of marine shrimp in Southeast Asia and China: major constrains for export. Southeast Asian Fisheries Development Center, Bangkok

FAO (2008) Biofuels: prospects, risks and opportunities. The State of Food and Agriculture 2008. Food and Agriculture Organization (FAO), Rome

FAO, NACA, UNEP, WB, WWF (2006) The international principles for responsible shrimp farming: shrimp farming and the environment. Network of Aquaculture Centres in Asia-Pacific (NACA), Bangkok

Flaherty M, Vandergeest P, Miller P (1999) Rice paddy or shrimp pond: tough decisions in rural Thailand. World Dev 27:2045–2060

Giap DH, Lebel L. (2006) An analysis of the shrimp aquaculture commodity chain between Thailand and the United States. USER Working Paper WP-2006-10. Unit for Social and Environmental Research: Chiang Mai

Goss J, Burch D, Rickson RE (2000) Agri-food restructuring and third world transnational: Thailand, the CP Group and the global shrimp industry. World Dev 28:513–530

Graslund S, Holmstrom K, Wahlstrom A (2003) A field survey of chemicals and biological products used in shrimp farming. Mar Pollut Bull 46:81–90

Holmstrom K, Graslund S, Wahlstrom A, Poungshompoo S, Bengtsson B-E, Kautsky N (2003) Antibiotic use in shrimp farming and implications for environmental impacts and human health. Int J Food Sci Technol 38:255–266

Hudson K, Watcharasakwet W (2007) The new Wal-Mart effect: cleaner Thai shrimp farms. Wall Street Journal, New York, 24 July 2007

Kasai C, Nitiratsuwan T, Baba O, Kurokura H (2005) Incentive for shifts in water management systems by shrimp culturists in southern Thailand. Fisher Sci 71:791–798

Keeratipipatpong W (2008) Shrimpers plan greener future. Bangkok Post, Bangkok 29 September 2008.

Kiatpathomchai S, Schmitz PM, BTS A, Thongrak S (2008) Investigating external effects of shrimp farming on rice farming in southern Thailand: a technical efficiency approach. Paddy Water Environ 6:319–326

Lebel L, Garden P, Luers A, Manuel-Navarrete D, Giap DH (2009a) Knowledge and innovation relationships in the shrimp industry in Thailand and Mexico. PNAS (in press)

Lebel L, Lebel P, Garden P, Giap DH, Khrutmuang S, Nakayama S (2008) Places, chains and plates: governing transitions in the shrimp aquaculture production-consumption system Globalizations 5:211–226

Lebel L, Mungkung R, Gheewala SH, Lebel P (2009b) From black to white: innovation cycles, niches and sustainability in the shrimp aquaculture industry in Thailand. USER Working Paper WP-2009-02 Unit for Social and Environmental Research, Chiang Mai University, Chiang Mai

Lebel L, Tri NH, Saengnoree A, Pasong S, Buatama U, Thoa Le Kim (2002) Industrial transformation and shrimp aquaculture in Thailand and Vietnam: pathways to ecological, social and economic sustainability? Ambio 31:311–323

Lee D, Connelly J (2006) Global Aquaculture Alliance on best aquaculture practices: an industry prepares for sustainable growth. Sustain Dev Law Policy 7:60–62

Lightner D (2005) Biosecurity in shrimp farming: pathogen exclusion through use of SPF stock and routine surveillance. J World Aquac Soc 36:229–248

MAP (2006) Wal-Mart and Darden Restaurants announce future sourcing of "certified" farm-raised shrimp: will consumers by served "green" shrimp or a green-wash? A position statement and action alert. Mangrove Action Project, Port Anglees, 29 January 2006

Mungkung R (2005) Shrimp aquaculture in Thailand: application of life cycle assessment to support sustainable development. University of Surrey, Surrey

NACA and FAO (2000) Aquaculture Development Beyond 2000: the Bangkok Declaration and Strategy. Conference on Aquaculture in the Third Millennium, 20-25 February 2000, Bangkok, Thailand. Network of Aquaculture Centres in Asia-Pacific and Food and Agricultural Organization, Bangkok and Rome

Naylor R, Goldberg R, Mooney H, Beveridge M, Clay J, Folke C, Kautsky N, Lubchenco J, Primavera J, Williams M (1998) Nature's subsidies to shrimp and salmon farming. Science 282:883–884

Naylor RL, Goldberg RJ, Primavera JH, Kautsky N, Beveridge MCM, Clay J, Folke C, Lubchenco J, Mooney H, Troell M (2000) Effect of aquaculture on world fish supplies. Nature 405:1017–1024

Neiland AE, Soley N, Varley JB, Whitmarsh DJ (2001) Shrimp aquaculture: economic perspectives for policy development. Mar Policy 25:265–279

Ongsritrakul S, Hubbard L (1996) The export market for Thai frozen shrimps in the European Union. Br Food J 98:24–28

Pananond P (2006) The changing dynamics of Thailand CP Group's international expansion. In: Suryadinata L (ed) Southeast Asia's Chinese businesses in an era of globalization: coping with the rise of China. Inst Southeast Asian Stud, Singapore, pp 321–359

Phillips M, Bueno P, Haylor G, Padiyar A (2003) From farm to plate: international trade in aquaculture products has a human development dimension of special interest to the Asia-Pacific region. Samudra 12–15

Menasveta P (2002) Improved shrimp growout systems for disease prevention and environmental sustainability in Asia. Rev Fisher Sci 10:391–402

Primavera J (2006) Overcoming the impacts of aquaculture on the coastal zone. Ocean Coastal Manag 49:531–545

QFFI (2003) Industry group defends shrimp farmers from attack by activists. Quick Frozen Foods Int 42:42

QFFI (2008) Global Aquaculture Alliance at odds with AFL-CIO shrimp labor study. Quick Frozen Foods Int 50:34

Rungfapaisarn K (2006) New IP unit to stimulate research projects. The Nation, 21 October 2006. Bangkok

Saengnoree A, Lebel L, Tri NH, Pasong S (2002) Technical barriers to trade, competitiveness and the sustainability of shrimp aquaculture in Thailand and Vietnam. The Conference of the Society for Ecological Economics, Sydney
Sagheer S, Yadav SS, Deshmukh SG (2007) Assessing international success and national competitive environment of shrimp industries of India and Thailand with Porter's diamond model and flexibility theory. Global J Flexible Syst Manag 8:31–43
Saowalak R (1999a) Shrimp farming and its sustainability: ecological footprint analysis (1). Prince of Songkla University J Environ Res 21(1):58–75
Saowalak R (1999b) Shrimp farming and its sustainability: ecological footprint analysis (2). Prince of Songkla University J Environ Res 21(2):51–75
Sapkota A, Sapkota AR, Kucharski M, Burke J, McKenzie S, Walker P, Lawrence R (2008) Aquaculture practices and potential human health risks: current knowledge and future priorities. Environ Int 34:1215–1226
Shimoda T, Fujioka Y, Srithong C, Aryuthaka C (2007) Effect of water exchange with mangrove enclosures based on nitrogen budget in *Penaeus monodon* aquaculture ponds. Fisher Sci 73:221–226
Center S (2008) The true cost of shrimp. Solidarity Center, Washington, DC
Stanley D (2000) The economics of the adoption of BMPs: the case of mariculture water management. Ecol Econ 35:145–155
Stonich SC, Bailey C (2000) Resisting the blue revolution: contending coalitions surrounding industrial shrimp farming. Hum Organ 59:23–36
Suwunnamek O (1997) The role of multinational agribusiness and the Shrimp Farmer's Cooperatives to the sustainability of the shrimp industry in Thailand. Agricultural Sciences University of Tsukuba, Ibaraki, Japan
Tacon AGJ, Forster IP (2003) Aquafeeds and the environment: policy implications. Aquaculture 226:181–189
Thai Rath (2008) Thai White shrimp number one in Asia [in Thai] 29 July 2008. Thai Rath, Bangkok
The Nation (2003) Funds sought for biotech development. The Nation. Bangkok, 19 April 2003
The Nation (2007) Shrimp farmers dispute Wal-Mart. The Nation, Bangkok 23 August 2007
Tanticharoen M, Flegel TW, Meerod W, Grudloyma U, Pisamai N (2008) Aquacultural biotechnology in Thailand: the case of the shrimp industry. Int. J. Biotechnology 10:588–603
Turakit Tanh Setakit (2004) Pooliang kung kaow pom ka yay vonokrom pramong raeng haa luk kung pon "wattana" nae rongrian hay kana korkor kaengkan [in Thai] 5 September 2004. Turakit Tanh Setakit, Bangkok
UNDP & FAO (1992) Terminal Report of the UNDP/FAO Project to establish the Network of Aquaculture Centers in Asia-Pacific. FAO Fisheries Department
Vaiphasa C, de Boer WF, Skidmore AK, Panitchart S, Vaiphasa T, Bamrongrugsa N, Santitamnont P (2007) Impact of solid shrimp pond waste materials on mangrove growth and mortality: a case study from Pak Phanang, Thailand. Hydrobiologia 591:47–57
Vandergeest P (2007) Certification and communities: alternatives for regulating the environmental and social impacts of shrimp farming. World Dev 35:1152–1171
Vandergeest P, Flaherty M, Miller P (1999) A political ecology of shrimp aquaculture in Thailand. Rural Sociol 64:573–596
World Bank, NACA, WWF FAO (2002) Shrimp farming and the environment: synthesis report. Work in progress for public discussion. World Bank, NACA, WWF, FAO, Washington, DC
Xinhua News Agency (2004) Shrimp research center to boost Thailand's competitive edge. Xinhua General News Service, 22 April 2004

Chapter 8
The Contribution of Organic Food Production to Sustainable Nutrition: A Case Study on the Organic Niche Market in Eastern Germany

Benjamin Nölting

Sustainability Problems Faced by the Food Sector

Healthy and tasty nutrition in sufficient quantities is a basic human need. The ways in which we eat and drink are part of human culture. At the same time, the production of food is closely linked with the natural environment and ecological cycles, demonstrating our dependence on nature. Altogether, food production and consumption form a complex system that is an integral part of our daily life.

Agriculture and food production are integrated into a multi-level system of markets, flows and actor networks. On a global scale, they have changed considerably over the last 50 years. Agricultural production and yields have been increasing due to the expansion of arable and irrigated land and the intensification, even industrialisation, of farming. International competition demands low prices, large volumes, standardisation, specialisation and high production efficiency from agricultural systems (Morgan et al. 2006). The globalisation of food markets has intensified long distance trade, resulting in complex commodity-flow and division of labour systems (cf. the shrimp industry, Chapter 7 in this volume).

As a consequence of these developments, many countries have increased their food security. In industrialised countries especially, an outstanding variety of food is available at low prices. On the other hand, agricultural intensification causes overexploitation and negative effects, such as soil degradation, water pollution, loss of biodiversity, climate change, and reduced standards of animal welfare (Knudsen et al. 2006). In 2009, about one billion people, mainly in developing countries, suffer from malnutrition, whereas in richer social sectors over-consumption and unhealthy diets are contributing to an increase of endemic illnesses, such as obesity or cardiovascular diseases. Food production and consumption are being exposed to structural changes, but these are not necessarily sustainable.

B. Nölting (✉)
Zentrum Technik und Gesellschaft (ZTG) der TU Berlin, Innovationsverbund Ostdeutschlandforschung, Berlin, Germany
e-mail: noelting@ztg.tu-berlin.de

L. Lebel et al. (eds.), *Sustainable Production Consumption Systems: Knowledge, Engagement and Practice*, DOI 10.1007/978-90-481-3090-0_8,

What are the challenges facing sustainable food production and consumption? To answer this question, we have to first define sustainable development. The research project "Regional Wealth Reconsidered",[1] on which this chapter is based, adopted the integrative approach of the Helmholtz-Gemeinschaft deutscher Forschungszentren (HGF), which defines three basic requirements for sustainable development: (a) ensuring human existence, (b) preserving the potential for production and reproduction, and (c) maintaining development potential. These requirements are implemented through 17 "rules" (Kopfmüller et al. 2001; Schäfer et al. 2004). According to this approach, we defined sustainable nutrition as being both environmentally friendly and healthy. Supply, information, and communication about sustainable nutrition should be in harmony with daily life routines and enables socio-cultural diversity. Sustainable nutrition should satisfy nutritional needs, while contributing to an improved quality of life (Eberle et al. 2006, pp. 53–54).

In this chapter we investigate the potential of organic agriculture and the organic food industry to foster the transformation towards sustainable food production and consumption. The focus is on Germany as an industrialised country, because industrialised countries such as those in the EU and the United States are a driving force in global trends due to their food markets and highly subsidised agriculture. In Germany, food production and consumption are constantly modernising in line with global trends. As a result, a complex system of food production has evolved that is increasingly efficient but is at least partly cut off from its natural and societal context. This creates sustainability problems in food production and nutrition, some of which have been analysed by, among others, a German network of social–ecological researchers focusing on "agriculture and nutrition" (Nölting and Schäfer 2007; Nölting 2008).

Intensive agricultural production is responsible for many environmental problems (UBA 1997, pp. 118–134). Food chains are becoming less and less transparent to consumers, and recurrent food scandals have reduced consumers' trust in food production, despite strict public standards with regard to food safety and increasing numbers of private labels and quality management systems (Eberle et al. 2006, pp. 67–72). The diversity of nutritional patterns has increased considerably: fast food, convenience food, and functional food are just some examples of this. In contrast, the nutritional skills required for ensuring healthy diets and home food preparation are partially eroding, with over-consumption consequently affecting consumers' health (Rösch and Heincke 2001). However, these consumption patterns can not be changed easily, because people have to manage nutrition in everyday lives that are subject to many constraints such as time, money, knowledge and accessibility (Eberle et al. 2006). Nutritional culture, to which consumers are often oblivious, is another important factor (Pfriem et al. 2006). Food production and consumption form a complex system; single interventions are not sufficient for solving sustainability problems. On the contrary, strategies need to address the whole food chain,

[1]The project "Regional Wealth Reconsidered" was funded under the Social–Ecological Research Programme by the German Ministry of Education and Research, from 2002 until 2007 (see: www.regionalerwohlstand.de).

including consumers; correspond with each other; and be synchronised in order to change production and consumption simultaneously (Nölting 2008; Brand 2008).

Against this background, the organic sector represents one important option, amongst others, for achieving a more sustainable agriculture and nutrition. This chapter deals with questions concerning what role the still relatively small organic sector can and should play in food production and consumption systems. Section on Organic Agriculture as a Sustainable Development Path in the Food Production Consumption System introduces organic agriculture as a possible path to sustainable development in the food production and consumption system and delineates a concept concerning how sustainable niches like organic food can influence conventional food regimes. In section on The Organic Agriculture and Food Industries in the Berlin-Brandenburg Region, a case study on northeastern Germany describes the organic sector's contributions towards sustainable development. The regional case study takes into consideration how food production and consumption systems are embedded in the national and regional societal and environmental contexts. Section on The Role of Knowledge: What do Consumers, Producers and Politicians (Need to) Know? analyses the knowledge required for change towards sustainability. In section on Action – Strategies for Expanding and Integrating the Organic Sector, strategies for the regional organic sector are developed, based on the premise that the regional level is an appropriate starting point from which to develop concrete strategies for action. The last section reflects on the potential outcome of regional strategies and asks whether the organic sector might be a model for more sustainable development.

Organic Agriculture as a Sustainable Development Path in the Food Production Consumption System

Food systems can be analysed using an approach normally used to analyse socio-technical systems, distinguishing three intertwined levels: (a) the socio-technical landscape, consisting of a set of deep structural trends; (b) socio-technical regimes, such as for food production and consumption, with rules and networks that coordinate activities in each field or the sector; and (c) technological niches. These niches are incorporated into their corresponding regimes, but are partially decoupled from them and "protected" from their rationale. They can thus generate even radical innovations and trigger regime alteration (Geels 2004a, 2004b). Introducing, supporting and establishing niches which contribute towards sustainable development may therefore be a plausible starting point for inducing change.

The organic agriculture and food sector – comprising agriculture, food processing, and food marketing – is seen as such a niche that might be able to overcome some of the above-mentioned problems (Belz 2004; Smith 2006). Other alternatives might include precision farming or fair trade (cf. Chapter 10 in this volume). The organic sector builds on a holistic understanding of the relationship between food production and nature and is characterised by an environmentally friendly mode of production (cf. Stolze et al. 2000; Stokstad 2002; Pimentel et al. 2005). The concepts underlying the practices of the organic agriculture sector are expressed by the principles of the International Federation of Organic Agriculture Movements (IFOAM) (IFOAM 2005):

- *Principle of health*: Sustain and enhance the health of soils, plants, animals and humans as being one and indivisible.
- *Principle of ecology*: Based on living ecological systems and cycles, work with them, emulate them and help sustain them.
- *Principle of fairness*: Build on relationships that ensure fairness with regard to the common environment and life opportunities.
- *Principle of care*: Managed in a precautionary and responsible manner to protect the health and well-being of current and future generations and the environment.

Organic farming started as a niche of food production. Due to continuous growth, however, the organic sector is now starting to infiltrate the regime of the conventional food sector. This development can be better understood by distinguishing between three different perspectives on what the organic sector is (or should be): (a) a protest movement; (b) a system with a common sensibility that continuously (re)creates itself through appropriate world views, principles, goals, and standards; and (c) a market niche (Alrøe and Noe 2008).

Organic farming started as an alternative social movement in Western Europe in the 1970s, introducing new ways of farming and deliberately establishing organic farming as a counter-model to conventional, intensive farming (Michelsen 2001). Since then the organic sector has evolved and become more sophisticated. The formal recognition of organic agriculture by the EU's Common Agricultural Policy in 1991 was a watershed moment in developing a common sensibility, because public institutions then became dominant actors in what had hitherto been a grassroots movement. Organic agriculture has been deemed to be in line with the EU's overall aims, because it minimises negative environmental effects, supplies high quality food, maintains income in the agricultural sector, and enhances rural development (Dabbert et al. 2004). The EU set standards for organic farming methods through a regulation on organic farming implemented in 1992 (Regulation EEC N° 2092/91) and one on production of animal species passed in 1999 (Regulation EC N° 1804/1999). These have been replaced by Council Regulation (EC) No 834/2007 of 28 June 2007 on organic production and labelling of organic products, which came into effect in January 2009, forming a comprehensive regulatory framework. According to these standards, organic farming should avoid the use of synthetic pesticides, herbicides, chemical fertilisers, growth hormones, antibiotics or genetic manipulation. The EU has also been providing financial assistance for organic farming under the umbrella of agro-environmental measures since 1992.[2] So far,

[2]For a detailed overview of the measures of the EU see Häring et al. 2004.

private and public actors have been developing rules for organic farming, so it has become an institution in the broad sense, with distinct norms for farm production (Michelsen et al. 2001, S. 5–6). While these organic institutions and standards are cornerstones of the common sensibility of organic agriculture, they are at the same time preconditions for the establishment of organic agriculture as a market niche.

Moreover, on a global scale organic agriculture has increased considerably over the past decade. In 2007, over 32 million hectares were managed organically in 141 countries. International sales of organic food have reached an estimated US$46 billion in 2007, with the organic food market being dominated by North America and Europe (Willer and Kilcher 2009). In some ways, organic agriculture is starting to grow out of its niche market status. In entering the conventional food market, the organic sector has had to adapt. Knudsen et al. (2006, pp. 28–34) have sketched two possible global trends for the future development of organic agriculture in industrialised countries:

1. A trend towards the "conventionalisation" of the organic sector, where it may become a slightly modified version of modern conventional agriculture
2. A counter-movement towards local production and consumption of organic food

Because of its lower yields, however, in developing countries, organic agriculture is probably not the key to overcoming hunger on a global scale. Yet, studies show that a considerable increase in organic agriculture in Europe and North America will not lead to a noticeable rise in world market prices and might even mitigate the distorting effects of overproduction and food dumping by the EU and the US. Moreover, organic agriculture as a non-certified, low-input technology could also contribute to self-sufficiency and improve local food security in developing countries (Halberg et al. 2006; El-Hage Scialabba 2007).

Taking all three perspectives together, the organic food sector (a) as a social movement provides alternative nutritional styles to consumers, (b) is recognised by policy makers and administrators as a stable and institutionalised system, and (c) is growing out of its niche-market status. We believe, therefore, that it can stimulate a more sustainable food production and consumption system, as will be demonstrated in the following case study.

The Organic Agriculture and Food Industries in the Berlin–Brandenburg Region

The organic agriculture and food sector is made up of farms that are certified according to EU standards, certified food processing firms, and organic food marketers and supporting organisations. In Germany, organic farming covered 5.1% of utilisable agricultural area in 2007 and organic food had a share of over 3% of the German food market, with a turnover of €5.3 billion in 2007 and high growth rates (Hamm et al. 2008).

Against this background, the results of a case study of the organic sector in the Berlin–Brandenburg are presented in the following paragraphs. The study's regional focus allows for an analysis of the broad range of economic, environmental, social and cultural effects of the organic sector. The social and cultural effects of the sector are often undervalued or even ignored. However, these aspects reflect the holistic nature of sustainable development and reveal both options for and constraints on structural change with regard to food consumption.

The Berlin–Brandenburg region is in northeastern Germany. It comprises the German capital Berlin (3.4 million inhabitants) and the surrounding federal state of Brandenburg (2.5 million inhabitants). Brandenburg is characterised by rural areas with low population density; little industrial infrastructure; with the exception of an industrialised belt around Berlin, a high unemployment rate of about 19%; and a fairly high percentage of state-protected natural preservation areas. Political, economic and social exchange between city and countryside is still limited due to the legacy of the Berlin Wall era, when Germany was divided into the Federal Republic of Germany (FRG) and the German Democratic Republic (GDR, the geographical area of which is now referred to as 'eastern Germany'). The wall cut off West Berlin from East Berlin and its surroundings for nearly 30 years. In 1990, German reunification set in motion a drastic transformation of the former East German regions within a very short period of time. Since that time, the number of employees in the agricultural sector of northeastern Germany has been reduced by 80%. In addition to economic problems, there is also demographic decline, with young people migrating away from rural areas, and state-run infrastructure is diminishing (Schäfer et al. 2009).

The organic sector had to start from scratch after German re-unification in 1990, but has been developing successfully in the Berlin–Brandenburg region since then. In 2007, 10.1% of the agricultural area in Brandenburg was cultivated organically (134,012 ha). In comparison to western Germany, a new type of organic farm has developed in the east, which is characterised by a greater average size (163 ha, compared to 31 ha in western Germany) and more specialised production. In addition, eastern products are often not sold solely within the region, but all over Germany. Because the conversion to organic production in the region is more a result of economic considerations than social protest, organic firms are not deeply rooted in the kinds of socio-cultural milieus typified by the environmental movement in western Germany.

The structure of the regional organic food production and consumption system is characterized by agricultural production (in Brandenburg) that tends to be sold outside the region. Meanwhile, Berlin is the decisive market for organic food in the region with an estimated sales of organic food of about €200 million and over 30 purely organic super markets; yet only about 10% to 20% of this market is supplied by regional organic products. The organic sector in eastern Germany was a latecomer and has developed within a specific institutional framework. This case is therefore of special interest, because it represents a new trend in the organic sector which is much more oriented towards conventional mass markets for food. What is the contribution of the organic sector in the Berlin–Brandenburg region to sustainable development?

Working from our definition of sustainable development, we applied its basic requirements and rules to the organic sector and the region. The research project developed criteria and indicators for its empirical analysis together with stakeholders from the organic sector in a participatory process. Empirical data was derived from a questionnaire based on this indicator set, which was sent to all (about 1,000) organic companies in the region in 2004, to which we received about 330 replies. Case studies were also carried out on selected enterprises to gain more insight into their activities (Schäfer and Illge 2006).

The empirical results show (Schäfer 2007) that the organic agriculture and food sector contributes to sustainable development by

- Contributing towards environmental protection and landscape aesthetics
- Preserving and creating social resources
- Passing on knowledge about dealing with nature and health issues

Environmental Protection and Landscape Aesthetics

Organic agriculture and food processing comply with EU organic standards. The ecological benefits include lower resource use, a significantly higher level of biological activity in the soil, higher levels of moisture retention capacity, and it is also the least detrimental farming system with respect to wildlife conservation and preserving landscape integrity. Our survey reveals that many organic firms even go beyond the EU standards and are committed to voluntary environmental activities. Farmers are taking measures to preserve the diversity of habitats and species such as planting hedges or trees, and installing wetlands (up to 60%); improving soil fertility (nearly 40%); and cultivating rare plants or animal species (nearly 30%). Of note is that they receive financial assistance only for some of these measures. In one case, a farmer even discussed the impact of his farm on the aesthetics of the local landscape with villagers.[3] Processing firms in this sector are more focused on saving energy and water or on establishing environmental management systems, such as an environmental audit.

Preservation and Creation of Social Resources

In demographically and economically declining peripheral rural areas, the stabilisation of social resources is very important. Over one third of the managers of organic companies are involved in local and regional associations (see Fig. 8.1). They participate

[3]Organic farming can be more favorable than land set-aside and succession, in particular on the poor soils in Brandenburg, because of its contribution to a diversified cultural landscape, with open fields benefiting both humanity and biodiversity.

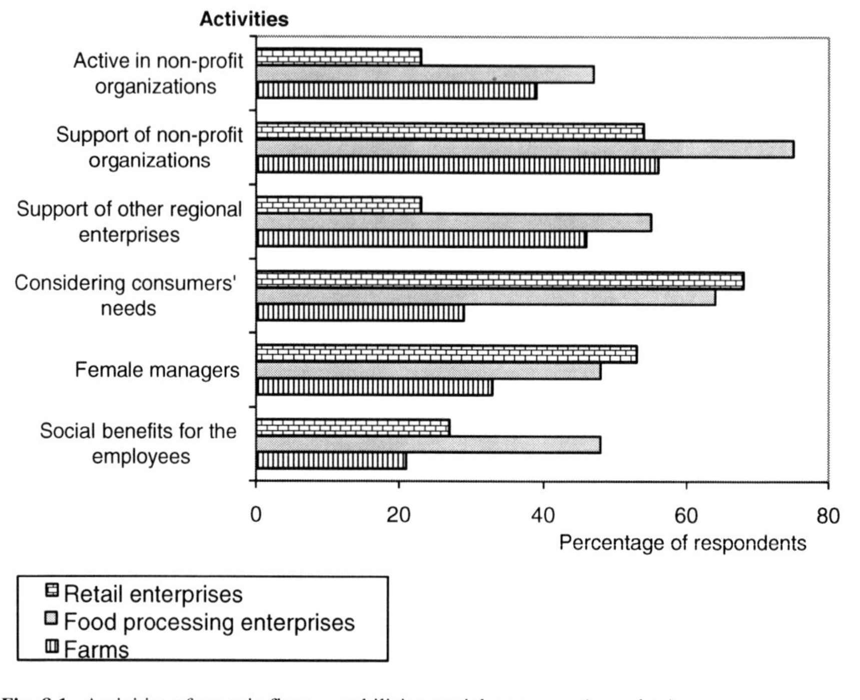

Fig. 8.1 Activities of organic firms – stabilising social resources (own data)

actively in environmental organisations, village clubs, regional networks such as the EU's Leader+programme, or in associations supporting large-scale natural protection areas. Even more of them support charity organisations by providing material or financial support. Involvement with natural parks and environmental or social organisations can often be a starting point for sustainable regional development projects. The specific goal of organic firms, namely the environmentally friendly production of high quality (regional) food, is a motivation for setting up these kinds of regional networks together with actors from the areas of natural protection, tourism, and trade.

Knowledge Transfer About Dealing with Nature and Health Issues

Knowledge transfer and the initiating of learning processes are essential for promoting and expanding sustainable lifestyles and modes of production. Formal knowledge transfer in vocational education on organic food production, processing, and marketing is only one aspect, however. Even more important for the transfer process is the

Fig. 8.2 Activities of organic firms – knowledge transfer (own data)

spread of information and communication in informal ways, such as organic firms providing open-house days, field trips or guided tours, participating in local festivities and events, providing flyers and websites, participating in public discussions concerning issues like GMOs, and retailers offering cooking courses (see Fig. 8.2). This kind of knowledge transfer combines cognitive learning with first hand experience. At a farm feast, city children might see a cow for the first time or, at a sales presentation, customers could learn how to prepare traditional types of vegetables. A cycling tour to an organic farm, a field walk or a visit to a dairy or bakery, followed by a meal in a farm café, can make a much deeper impression than dozens of leaflets, websites and lectures. Such personal experience often seems to be more authentic and meaningful, in great part due to the conviction and commitment of the people working in the organic sector, and better stimulates consumers' learning processes.

To sum up, the results show that enterprises in the organic sector are committed to a variety of activities beyond an environmentally friendly mode of production. Organic enterprises preserve the natural environment, stabilize social resources, and pass on knowledge about environmentally friendly agriculture and food production and healthy nutrition. Organic firms are often a nucleus for further sustainability initiatives, as our case studies in Berlin–Brandenburg show. Further, the organic sector also represents a sustainable means of nutrition and lifestyle. Organic food has become an additional option for consumers in the region, even though they still mainly purchase conventional food. The manifold character of these sustainability activities underlines that organic actors are not restricted to an economic rationality, but that they are interested in developing social, cultural, and political relationships with consumers and society. These links have strong roots in regional exchanges and networks, triggering learning processes concerning sustainable nutrition.

The organic agriculture and food sector can thus be regarded as a sustainable production system in the region. The results reveal the organic sector's high potential for making the food production and consumption system more sustainable, because it contributes to the transition strategies suggested by Lebel and Lorek (see Chapter 14): slowing down food production and consumption in cases of over-production and over-consumption, making the benefits and deficits of conventional and organic food more 'visible', and linking knowledge about food with concrete action and activities.

The Role of Knowledge: What Do Consumers, Producers and Politicians (Need to) Know?

The "technology" of the organic food production and consumption system is based on knowledge, not on material inputs. Organisational, technological, political, and cultural innovations are the main drivers of this sustainability transition. Three different groups have stakes in the organic sector: consumers, producers, as well as politicians and the public setting framework conditions for production and consumption. Their knowledge is one factor in the transition towards sustainability.

Knowledge of Consumers

Consumers do not decide about sustainable nutrition only at the checkout counter, weighing up prices and information. In fact, they are bound to daily routines, financial restrictions and nutritional culture. Consumers also feel insecure about and rather patronised by the huge quantity of information on the "right" nutrition that has flooded their lives in the past years, so consumer communication has to be made from a consumer's perspective (Hayn 2007; Büning-Fesel 2007). As described above, many organic firms pass on knowledge and experience about sustainable agriculture, healthy nutrition, and tasty food to consumers in different ways. An important reason for doing so is that they have to explain the specific qualities – and the higher prices – of organic food to customers because some of the qualities of organic food are not inherent in the food itself, but rather in production processes such as ensuring a high level of environmental quality and animal welfare. Even so, organic firms provide knowledge which goes far beyond the product information required for marketing purposes, providing examples of sustainable lifestyles and initiating learning processes (Boeckmann 2009). Close and trusting communication between producers and consumers can have a deep impact, but it is costly and time-consuming. Labels and production standards also play an important role in disseminating information to a large number of consumers. Transparent food chains are an important element of arguments in support of organic food. Indeed, organic firms are, partly for economic reasons, involved in taking measures to promote sustainable consumption, such as codes and standards, consumer education and the evaluation of "simple" food and nutritional styles.

Norms and values are closely related to knowledge and play a crucial role in consumers' decisions as to what they want and buy. The organic sector presents alternative values, discusses its environmental and social responsibility, and promotes a guiding vision for sustainable food production and consumption. Although it has not been supported by the majority of consumers, it raises questions and initiates public debate. These activities often go beyond the reach of organic firms. Such normative debates need to be embedded in the broader social context of social movements and political discourses. In addition to information and values, however, other factors such as price, accessibility, and routines are crucial for (sustainable) consumption.

Knowledge of Producers

Organic farmers and food processors need more technologically related knowledge in order to improve environmentally friendly food production techniques, for example large-scale agriculture or farming adapted to regional conditions (dry climate, sandy soils). In fact, research on organic food production has been promoted by the German Ministry of Food, Agriculture and Consumer Protection in recent years. But business opportunities, market structure, regulations, and financial and political support are more important factors than technological knowledge and market information for producers.

Knowledge of Politicians and in Public Discourses

Agricultural policy has been quite effective in expanding organic agriculture through financial assistance, and organic standards have been implemented through government regulation and control (Häring et al. 2004; Dabbert et al. 2004). This type of regulatory approach, however, does not stimulate innovation and cooperation. Regional policy makers need "systemic" knowledge on how to initiate and manage cooperation between the organic producers, consumers, citizens, politics and administration in order to foster relationships between Brandenburg's rural areas and the city of Berlin. However, political support for the regional organic food production and consumption system depends more on political priorities and values and less on knowledge.

Stakeholders and experts from the field have very detailed and specific knowledge that could promote sustainable transition. However, sustainability communication could still be further improved by linking consumers, producers, and policy makers more closely. The plurality of knowledge, motives, norms and ideals provided by actors in the organic sector are a very important asset for stimulating innovation, fueling public debates about what kind of agriculture and food society wants, and initiating learning processes.

Action – Strategies for Expanding and Integrating the Organic Sector

Knowledge is one essential aspect for promoting the organic sector. Economic incentives, market opportunities, policy regulation, and consumers' wants are other crucial determinants. What are appropriate strategies for expanding the organic sector in Berlin–Brandenburg – and by implication possibly other regions – beyond the niche market? The challenge is to transfer the sustainability innovations of the organic niche to the dominant food regime. There seems to be a trade-off between a deep sustainability impact and increasing the share of organic food at the conventional food market, because the organic sector will have to restructure its production according to the logic of conventional mass markets when it tries to enter them (Lockie and Halpin 2005; Smith 2006). The organic niche has to be innovative and trustworthy as well as compatible with mainstream consumers and market structures. As a consequence, any strategy for action has to foster the quantitative expansion of organic food production (broadening) as well as its qualitative impulses and innovations towards sustainable nutrition (deepening). A single strategy will not be sufficient.

As described above, the organic sector in Berlin–Brandenburg is integrated more into national markets than into the economic and social context of the region. Three economic, social and political integration strategies are therefore suggested to strengthen regional and societal links, which tend to be more stable than market trends. These strategies have to correspond in order to enhance their impact, because the dominant food regime relies on self-stabilising structures and mechanisms which are difficult to overcome (Geels 2004a).

Economic Integration

Demand for organic food was and still is rising rapidly, especially in Berlin, and demand for regional organic products has started to develop. This provides an opportunity for organic enterprises to expand regional value-added chains in order to provide the regional market with organic food and open up new markets. Because of the small quantities and limited variety available, however, the marketing of regional organic food there has been difficult so far. But, for increasing numbers of suppliers, especially for small and medium size companies and processing firms, the regional market is an interesting opportunity which can create production synergies with national markets, all the more so as the organic food market in Berlin has reached a critical mass. Organic foods from regional value-added chains can supplement national and international food chains and expand the range of choices available to consumers who want to support farmers from the region or avoid food miles. The advantage of regional food is that the production chains can be easily made clear to consumers and underpinned by direct contact, by visits and regional networks. It still lacks economies of scale however and is, thus, probably not suitable for conquering mass markets.

Social Integration

Close communication with consumers helps producers to respond smoothly to their wishes. On the other hand, a campaign for regional organic food may be a promising way for communicating about and promoting sustainable food consumption. Such a campaign has to make all contributions to sustainable development of the organic sector's 'visible'. A good example is the initiative "fair & regional" for organic products from farmers, processing firms and trading enterprises in Berlin and Brandenburg (www.fair-regional.de). Any such communication strategy, however, has to go beyond marketing and contribute to public discourses about sustainable lifestyles and food consumption. Activities such as knowledge transfers, supporting socio-cultural institutions, forming alliances with other stakeholders from environmental organisations, or being involved in consumer protection and regional development, etc. foster social interaction between the organic sector and the broader society. Such interaction and communication campaigns require the cooperation of many organic firms and organisations. Third party involvement is also necessary, and policy is a key player in this respect.

Political Integration

The organic sector in the Berlin–Brandenburg region offers great potential for supporting and even initiating sustainable development processes. As a consequence, policy makers should support the organic sector through an organic action plan, and not only because it is a flourishing economic sector. They should also address it as a partner in sustainable regional development, with organic firms becoming actively involved in local and regional networks. A broad strategy of sustainable regional development should attempt to deepen the relationship between city and countryside. The organic sector is rooted not only in rural areas, but also in the Berlin market, so it can combine complementary aspects – such as the natural beauty of the countryside, the production of high quality organic food – with the city's broad range of economic, social, and cultural choices. Any policy strategy and political support for organic food production and consumption should, for this reason, include organic farmers, processing firms and trading firms as well as consumers' associations, the health sector, environmental organisations, and educational institutions. Policies could for example promote public organic food procurement with regional products and combine it with educational campaigns in schools and kindergartens. Political support could stabilise the organic niche, with its positive environmental and social effects, and even strengthen its power to transform the current food production and consumption regime.

Conclusion: Organic Food Production – a Model for Sustainable Agriculture and Nutrition

Can the organic sector in the Berlin–Brandenburg region be a model for food production and consumption in general? Can it become a mainstream concept within the new logic of the socio-technical regime of food production? Or will it be restricted to remaining a niche market? The organic food market tends to differentiate mass production for the market from producing high-quality food and more expensive specialities. The challenge is therefore a double one: on the one hand, the organic sector has to gain a higher market share to expand its impact (broadening) while, on the other hand, it has to maintain its vitality and combine the elements of protest, a particular sensibility regarding the connection between nature and society and high-quality products in order to deepen sustainable food production and supply (Alrøe and Noe 2008).

Changing the food system in industrialised countries is important in order to minimise the detrimental effects of over-consumption and over-production. The organic food sector provides "green" products and alternative, sustainable nutrition. It demonstrates how sustainable food production can work, establishes trusting producer–consumer relationships, stimulates learning processes with colleagues and consumers, and initiates debates about values and norms with regard to sustainable food and nutrition (fair prices, fair trade, GMO, healthy, environmentally friendly nutrition etc.) Due to its plurality, the organic sector is a driving force for sustainable transition, qualitatively and quantitatively, and its example is challenging the conventional food sector. Of course, it can be and needs to be optimised and expanded in order to overcome the restrictions typical of niche markets. The regional scale is not sufficient to overcome all the sustainability problems of the food sector, but it highlights the importance of societal integration of the food industry. To conclude, the organic sector indicates a path towards a sustainable food production and consumption system on both the material as well as the symbolic levels.

References

Alrøe HF, Noe E (2008) What makes organic agriculture move – protest, meaning or market? A polyocular approach to the dynamics and governance of organic agriculture. Int J Agric Res Govern Ecol (IJARGE) 7(1/2):5–22

Belz FM (2004) A transition towards sustainability in the Swiss agri-food chain (1970–2000): using and improving the multi-level perspective. In: Elzen B, Geels FW, Green K (eds) System innovation and the transition to sustainability. Edward Elgar Publishing, Northampton, MA, pp 97–113

Boeckmann T (2009), Biohöfe als Ausgangspunkte informeller Lernprozesse. Eine explorative Studie über die Zusammenhänge zwischen dem Handeln von ökologisch wirtschaftenden Landwirtschaftsbetrieben und nachhaltigen Lebensstilen und Wirtschaftsweisen im Hofumfeld. Kassel/Witzenhausen, Diss.

Brand KW (2008) Konsum im Kontext. Der "verantwortliche Konsument" – Ein Motor nachhaltigen Konsums? In: Lange H (ed) Nachhaltigkeit als radikaler Wandel. Die Quadratur des Kreises? VS Verlag für Sozialwissenschaften, Wiesbaden, pp 71–94

Büning-Fesel M (2007) Vermittlung und Umsetzung eines nachhaltigen Ernährungsstils – Anregungen aus Sicht des aid infodienst. In: Nölting B, Schäfer M (eds) Vom Acker auf den Teller. Impulse der Agrar- und Ernährungsforschung für eine nachhaltige Entwicklung. Oekom Verlag, München, pp 165–172

Dabbert S, Häring A, Zanoli R (2004) Organic farming: policy and prospects. Zed Books, London

Eberle U, Hayn D, Rehaag R, Simshäuser U (eds) (2006) Ernährungswende. Eine Herausforderung für Politik, Unternehmen und Gesellschaft. Oekom Verlag, München

El-Hage Scialabba N (2007) Organic agriculture and food security. FAO, Rome

Geels FW (2004a) From sectoral systems of innovations to socio-technical systems: insights about dynamics and change from sociology and institutional theory. Res Policy 33:897–920

Geels FW (2004b) Understanding system innovations: a critical literature review and a conceptual synthesis. In: Elzen B, Geels FW, Green K (eds) System innovation and the transition to sustainability. Edward Elgar Publishing, Northampton, MA, pp 19–47

Halberg N Sulser TB, Hogh-Jensen H, Rosegrant MW, Knudsen MT (2006) The impact of organic farming on food security in a regional and global perspective. In: Halberg N, Alrøe HF, Knudsen MT, Kristensen ES (eds) Global development of organic agriculture: challenges and promises. Cabi Publishing, Wallingford, pp 277–322

Hamm U, Liebing R, Richter T (2008) Bio sucht Bauer!. Ökologie & Landbau 36(3):14–17

Häring AM, Dabbert S, Aurbacher J, Bichler B, Eichert C, Gambelli D, Lampkin N, Offermann F, Olmos S, Tuson J, Zanoli R (2004) Organic farming and measures of European agricultural policy. Universität Hohenheim, Stuttgart-Hohenheim

Hayn D (2007) Alltagsgestaltung der Konsumentinnen und Konsumenten – Ausgangspunkt einer Ernährungswende. In: Nölting B, Schäfer M (eds) Vom Acker auf den Teller. Impulse der Agrar- und Ernährungsforschung für eine nachhaltige Entwicklung. Oekom Verlag, München, pp 73–83

IFOAM (2005) Principles of organic agriculture. IFOAM, Bonn

Knudsen MT, Halberg N, Olesen JE, Byrne J, Iyer V, Toly N (2006) Global trends in agriculture and food systems. In: Halberg N, Alrøe HF, Knudsen MT, Kristensen ES (eds) Global development of organic agriculture: challenges and promises. Cabi Publishing, Wallingford, pp 1–48

Kopfmüller J, Brandl V, Jörissen J, Paetau M, Banse G, Coenen R, Grunwald A (2001) Nachhaltige Entwicklung integrativ betrachtet. Konstitutive Elemente, Regeln, Indikatoren. Edition Sigma, Berlin

Lockie S, Halpin D (2005) The 'Conventionalisation' thesis reconsidered: structural and ideological transformation of Australian organic agriculture. Sociol Ruralis 45(4):284–307

Michelsen J (2001) Organic farming in a regulatory perspective: the Danish case. Sociol Ruralis 41(1):62–84
Michelsen J, Lynggaard K, Padel S, Foster C (2001) Organic farming development and agricultural institutions in Europe: a study of six countries. Universität Hohenheim, Stuttgart
Morgan K, Marsden T, Murdoch J (2006) Worlds of food: place, power, and provenance in the food chain. Oxford University Press, Oxford
Nölting B (2008) Social–ecological research for sustainable agriculture and nutrition. ZTG discussion paper, Zentrum Technik und Gesellschaft der TU Berlin, Berlin, 27 Aug 2008
Nölting B, Schäfer M (eds) (2007) Vom Acker auf den Teller. Impulse der Agrar- und Ernährungsforschung für eine nachhaltige Entwicklung. Oekom Verlag, München
Pfriem R, Raabe T, Spiller A (eds) (2006) OSSENA – Das Unternehmen nachhaltige Ernährungskultur. Metropolis-Verlag, Marburg
Pimentel D, Hepperly P, Hanson J, Douds D, Seidel R (2005) Environmental, energetic, and economic comparisons of organic and conventional farming systems. BioScience 55(7):573–582
Rösch C, Heincke M (2001) Ernährung und Landwirtschaft. In: Grunwald A, Coenen R, Nitsch J, Sydow A, Wiedemann P (eds) Forschungswerkstatt Nachhaltigkeit. Edition Sigma, Berlin, pp 241–263
Schäfer M (ed) (2007) Zukunftsfähiger Wohlstand. Der Beitrag der ökologischen Land-und Ernährungswirtschaft zu Lebensqualität und nachhaltiger Ernährung. Metropolis, Marburg
Schäfer M, Illge L (2006) Analyzing sustainable wealth – the societal contributions of a regional industrial sector. In: Estes RJ (ed) Advancing quality of life in a turbulent world. Springer, Dordrecht
Schäfer M, Nölting B, Engel A (2009) Organic agriculture as a new player in sustainable regional development? Case studies of rural areas in Eastern Germany. Int J Agric Res Govern Ecol (IJARGE) 8 (2/3/4): 158–179
Schäfer M, Nölting B, Illge L (2004) Bringing together the concepts of quality of life and sustainability. In: Glatzer W, Below Sv, Stoffregen M (eds) Challenges for the quality of life in contemporary societies. Kluwer, New York, pp 33–43
Smith A (2006) Green niches in sustainable development: the case of organic food in the United Kingdom. Environ Plann C: Govern Policy 24:439–458
Stokstad E (2002) Organic farms reap many benefits. Science 296 (May):1589
Stolze M, Piorr A, Häring A, Dabbert S (2000) The environmental impacts of organic farming in Europe. University of Hohenheim, Stuttgart
UBA, Umweltbundesamt (1997) Nachhaltiges Deutschland. Wege zu einer dauerhaft umweltgerechten Entwicklung. Erich Schmidt, Berlin
Willer H, Kilcher L (eds) (2009) The world of organic agriculture: statistics and emerging trends 2009. Frick, Genf, IFOAM, FiBL, ITC, Bonn

Chapter 9
Sustainable Information in the Pork Chain

Agni Kalfagianni

Introduction

Agricultural and food objectives have shifted during the past decades from ensuring food sufficiency to striving for food sustainability. At the end of the Second World War, a devastated Europe searched for ways to feed its undernourished population. In the European political agenda at that time, issues of food security, land reforms, increasing productivity and technological improvement scored very high. The aim was to produce enough affordable food for society. At the national levels, state-driven policies supporting the industrialization, intensification and rationalization of agricultural production were put forward with the adoption of the Fordist model of increasing wage/productivity (McMichael 1997) through American led reconstruction programs such as the Marshal Aid (Goodman and Redclift 1991; Marsden et al. 1996; Ward and Almas 1997). At the same time, industrialization – which paid much higher wages than labour in agriculture – occurred in different sectors of the economy and resulted in urbanization and rural exodus. For that reason, subsidies were introduced to keep agricultural labour from lapsing to competitive fields and secure production. The results were rewarding: agriculture began to transform from a relatively backward and highly labour-intensive sector of the economy towards one of increasing technological sophistication (Bowler 1985; Gardner 1996), while the process of business termination slowed down (Van Leeuwen, 2003).

The development of the Common Agricultural Policy (CAP) in 1957 along with the Treaty of Rome and the establishment of the European Economic Community (EEC) further promoted and harmonized the national objectives set for agriculture. State intervention and price support schemes were further promoted to secure an

A. Kalfagianni
Department of International Relations and European Integration, University of Stuttgart, Stuttgart, Germany
e-mail: agni.kalfagianni@sowi.uni-stuttgart.de

L. Lebel et al. (eds.), *Sustainable Production Consumption Systems: Knowledge, Engagement and Practice,* DOI 10.1007/978-90-481-3090-0_9,

income for the farmers[1] and adequate food for society. Moreover, a reduction of barriers to trade between the EEC member-states was introduced and common prices for agricultural products were set. As a result, the CAP and national policies achieved self-sufficiency in food, stability in agricultural markets and a fair standard of living for the farmers in Western Europe.

Although CAP and national policies were successful in their objectives, they created a number of problems which shifted the aims and operation of subsequent agricultural policies. Specifically, the intensive forms of production promoted by the European Union CAP and national policies have had severe consequences both for the environment and human health.

Knowledge on Unsustainable Impacts of the Agricultural Production and Consumption System

The agricultural sector in particular, has proven to be an important source of air pollution (e.g. CH_4 from cattle farming, waste and animal husbandry; N_2O from the use of synthetic nitrogen fertilizers), and greenhouse gas emissions (CO_2 resulting from the use of fossil fuel and the production of agricultural inputs) (Biesiot and Moll 1995), contributing to global warming, acidification and eutrophication and causing health problems. Moreover, studies have established that emissions increase as the intensity and scale of agricultural production amplify (Kramer et al. 1999). The rapid industrialization and intensification of agriculture has also been responsible for the continuing decline of biodiversity in agricultural landscapes, a trend observed throughout Europe (Andreasen et al. 1996; Baldock 1990; Delbeare et al. 1998; Fuller et al. 1995; Manhoudt and de Snoo 2003). Intensive agriculture is considered responsible for the extensive drainage and extraction of groundwater, causing groundwater shortages, decline of groundwater-dependent ecosystems and poor water quality (Van Ek et al. 2000). Similarly, the intensive use of agricultural land affects the long-term production capacity of the soil, which is crucial for a continued supply of high quality foodstuffs.

In addition to the agricultural sector, the stages of processing, packaging, storing and transportation have also been significant in terms of their impact. In meat production, studies report that the processing stage causes the largest environmental impact due to production of water effluents with high organic waste content. This kind of waste is very difficult to purify and dispose of because it is predominantly made from wastewater coming from all stages of the meat production process,

[1]These schemes were based on the establishment of (high) target market prices for agricultural products and the setting up of lower intervention prices to account for the potential failure of the market to meet the target prices. Specifically, the intervention schemes worked as follows. The Commission set a target price for the agricultural products, which was supposed to be met by demand and supply in the market. If, however, the market did not support the target price, then the Commission started to buying the product itself at the intervention price.

including washing, cleaning, scalding as well as from the water boilers and cooling machinery (UNEP 2000). Similar observations are made for the production of fish. Fish production is reported to contribute even more to waste because of its high perishable nature in comparison to other foods, and the associated large losses that occur during the production chain as a whole (UNEP 2000).

Moreover, intensive animal production methods reportedly cause important human health and safety hazards such as joint, kidney, and heart problems (Buzby 2002), infections (Tauxe 2002), various kinds of cancer (Hill 1999; Lijinski 1999; McKnight et al. 1999; Navarro et al. 2003; Norat et al. 2002; Peters et al. 1992; Willet et al. 1990, 1996) and even diseases that are thought to be extinct from Western countries such as hepatitis E (Hoekstra 2002; Van der Poel et al. 2001). More dramatically, in terms of concentrated effects in a short period of time, intensive animal production methods also foster the outbreak of assorted animal diseases, such as pig plague, swine fever, salmonella, and Bovine Spongiform Encephalopathy (BSE). After the initial outbreak of BSE, society was shocked not only by the revelation of the fact that one could actually die by eating meat but also by the way animals were treated. Consumers began to question the ability of the modern food system to provide safe food (Smith and Riethmuller 2000; Tansey and Worsley 1995; Yeung and Morris 2001) and called for more attention to environmental and health problems as well as animal welfare concerns.

As a result of increasing knowledge a shift in policy objectives regarding agriculture and food took place. The concept of sustainability and sustainable development was gradually introduced as a core element of national and regional (EU) policies. Today agricultural and food policies in pursuit of sustainable development must consider environmental and social consequences, in addition to economic and food security concerns. Policy makers realized that agricultural and food policies should not only concentrate on securing an income for producers and sufficient food for society but also must take into account environmental and health aspects. This is illustrated, for instance, in the two reforms of the CAP (1992, 2000) which aimed at the adoption of measures that encourage "farming practices compatible with the increasing demands of protection of the environment and natural resources and upkeep of the landscape and the countryside" (Council of the European Communities 1992). In addition, the adoption of a number of directives in the area of environmental policy that supplement some of the provisions of the reforms are also illustrative of the (at least formally stated) shift that is being realized in agricultural and food policies. Important examples include the Directive 91/676/EEC concerning the protection of wastes against pollution caused by nitrates from agricultural sources aiming to limit the spreading of fertilizer containing nitrogen and to set the limits for the spreading of livestock effluent; the water framework (1999) which sets out to achieve good water status for all waters by 2015; the Integrated Pollution Prevention and Control (IPCC) Directive with the aim to prevent or minimize emissions to air, water and soil, as well as waste from industrial and agricultural installations in the community; and the Pesticide Directive (1991/414) concerning the placement of plant protection products in the market.

Knowledge Provision along the Food Chain

The promotion of the sustainability objectives set for agriculture and food however, can only be successful if policies for the establishment of transparency in the food chain are promoted simultaneously. Transparency is conceptualized in two dimensions: one vertical, referring to the tracking and tracing of the meat product chain stages, and one horizontal, identifying the impacts on sustainability from the various stages in the chain. A high degree of transparency in the vertical dimension is important in order to ensure the accurate and rapid identification of product and process information up and down the chain. This process is also known as *traceability* and is primarily associated with food safety. A wide scope of transparency in the horizontal dimension on the other hand, ensures that the full chain impact on sustainability can be estimated and judged. In addition this impact becomes visible to all the actors and society at large therefore enabling interventions for sustainability. Together, the degree and scope of transparency in the vertical and horizontal dimensions define the overall level of transparency in the chain. In terms of comprehensiveness and need for the promotion of sustainability in the meat chain, the desirable level of transparency in the chain is established when the vertical degree and horizontal scope of transparency approach a maximum. A maximum degree of transparency in the vertical dimension is achieved when all the stages of the meat product chain are required to be traced by the policy, and a maximum scope of transparency in the horizontal dimension is achieved when all aspects related to sustainability (see Fig. 9.1) are required to be covered by the policy.

This paper argues that transparency is fundamental for shedding light on the sustainability aspects of food production and consumption, in particular aspects concerning human and animal health and safety, animal welfare and the environment. With transparency, the conditions under which the modern food system operates become tangible and make interventions in favor of sustainability possible. From a consumer perspective too, transparency helps in minimizing the spatial and cultural distance between production and consumption created by globalization and trade liberalization, which currently prohibits actors in later stages of the chain (including final consumers) from using reliable sustainability criteria in their consumption choices. The establishment of food chain transparency depends however, on all the links in the chain, from the production of feed ingredients to the final product. This requires actors forming the chain links (food chain actors), such as farmers, processors and retailers, to make sustainability related information available for its communication in the chain and towards final consumers. This chapter seeks to assess the level of transparency in the pork chain in the Netherlands and the EU, and identify obstacles and opportunities for its improvement.

The case selection was made on the basis of the following criteria. Pork production chains are highly intensive in terms of consumption of environmental resources and generation of waste, while industrialized production methods also raise animal welfare concerns. At the same time pork is the most important meat product sector in terms of revenues in the Netherlands, thus its economic significance for the

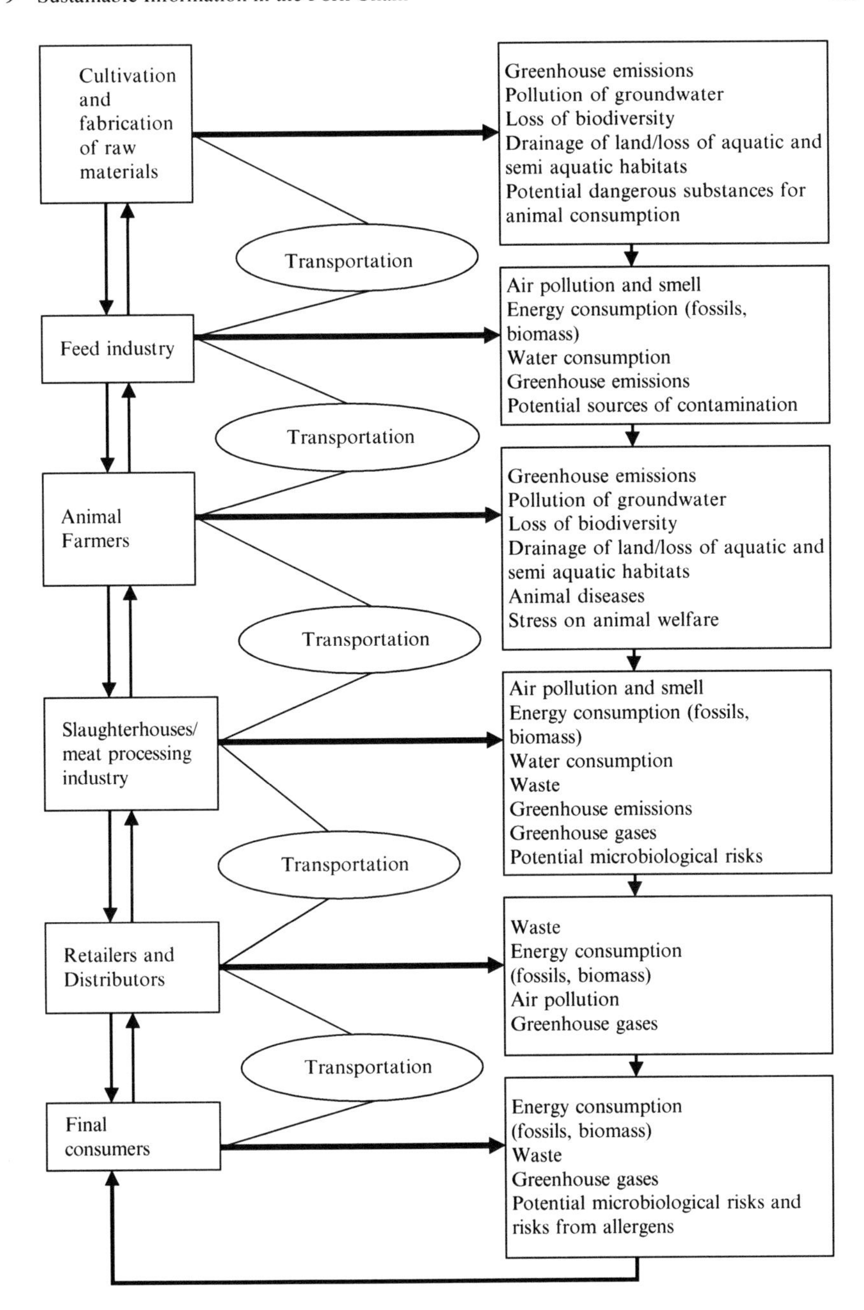

Fig. 9.1 Transparency in the commodity chain for meat and meat products (adapted from Kalfagianni 2006:21)

Dutch agricultural and food sector is very large. Pork is important at the EU level as well especially from an export value perspective, as the EU is the second largest pork exporter in the world after China.

The rest of the paper is organized into the following sections: the detailed status quo with respect to the state of sustainability related information in the pork chain at the EU level and Dutch level at present, obstacles for an improvement of the status quo and the conclusion.

Transparency in the Pork Production and Consumption System: Initiatives at the EU Level

The EU took the first step towards establishing a system of distribution of information regarding foodstuffs among the actors in the food chain as a whole after the BSE crisis in 1996. On 30 April 1997, the European Commission published a Green Paper on European Food Law with the intention of launching a public debate on the need to change the existing legislation on foodstuffs. The Commission aimed to meet the expectations of all the parties involved in the supply chain and to ensure that control and inspection systems meet the objectives to ensure a wholesome supply. The Commission stressed the need to provide a central unifying text setting out the fundamental principles on food law and clearly defining the obligations of the parties concerned. For the first time the need for a "from farm to table" approach on the regulatory framework for foodstuffs was realized as a means to ensure food safety and re-establish consumer trust.

On 12 January 2000, in the wake of a new dioxin crisis in Belgium,[2] the European Commission adopted a White Paper on Food Safety based on the consultation arising from the Green Paper on Food Law. The White Paper stressed the need for an integrated approach after recognizing that production chains have become both extremely complex and integrated within the single market. This means that development in farming and food process and distribution patterns could not be dealt with sectorally, as had been the case. Therefore, a food safety policy would only be effective if it acknowledged the inter-linked nature of food production. For this reason, the White Paper emphasized that the role of all stakeholders in the food chain (the food chain actors; the competent authorities in Member States and third countries; the Commission, and the consumers) should be clearly defined. According to the White Paper, the primary responsibility for food safety should lie with the food chain actors who would ensure that adequate procedures are in place to withdraw food and feed from the market when it poses a risk to the health of the consumer.

In December 2000, after a long period of consultation with all the relevant stakeholders (including consumer groups and third country authorities) the Commission published a proposal for a Regulation on Food Law of the European Parliament and the Council. In January 2002, the European Parliament and the Council adopted

[2]This event had very large impact not only for the improvement of transparency but also for the EU food safety policies in general, as well as the re-organization of the EU bodies responsible for food.

that regulation (178/2002/EC) (applicable from 1 January 2005). The regulation imposes concrete demands for *traceability* from food chain actors alongside the chain (excluding final consumers). According to Article 18 of the regulation "traceability means the ability to trace and follow a food, feed, food producing animal or substance intended to be, or expected to be incorporated into a food or feed, through all stages of production, processing and distribution". Article 18 states the requirements for traceability: specifically that food and feed business operators should be able to identify any person from whom they have been supplied with a food, a feed, a food-producing animal, or any substance intended to be or expected to be incorporated into a food or feed. To this end such operators should have systems and procedures in place allowing for the on demand availability of this information to competent authorities. In addition, food and feed business operators should have in place systems and procedures identifying other businesses to which their products have been supplied. This information should also be made available to competent authorities on demand. It is required that food or feed placed on the market or likely to be placed on the market in the Community should be adequately labeled or identified to facilitate its traceability through relevant documentation or information in accordance with the relevant requirements or more specific provisions. It is also stated that traceability needs to be ensured at all stages including third countries. Traceability then, at its current form, is an informational tool aiming to ensure food safety. As such, the principal instrument for the distribution of information among actors in the food chain is related, in fact, to the vertical dimension of transparency and covers a rather narrow space within the spectrum of sustainability aspects.

In addition, to traceability, an EU initiative for the entire chain, other policies and initiatives for the distribution of information have been developed both at the EU, targeting specific sectors, or developed from sector actors for the chain. These are described in some detail below.

Feed Sector

The feed sector is organised into the *compound feed sector* and the *ingredients feed sector.* The compound feed sector represents one third of the total feed consumed every year (140 million tonnes out of the total 450 million tonnes that EU animals consume every year) while the other two thirds represent ingredients that can be produced by the farmers themselves at the farm. Pigs and chickens are almost entirely fed with compound feed while cattle and sheep are fed with feed ingredients.

For the feed sector *safety* is a primary concern as many food related crises begin with feed mishandling. As such, the instruments for the distribution of information described here focus on ensuring safety, again, rather than promoting general sustainability objectives. To guarantee feed safety, the sector is regulated by the feed hygiene regulation which endorses the use of the Hazard Analysis of Critical

Control Points (HACCP) system. Until recently, this regulation only included the compound feed industry exempting home mixers and producers of feed ingredients. However since December 2004 a new EU feed hygiene regulation is in place (applicable from 1 January 2006), covering the use of HACCP for all feed business operators including the producers of feed ingredients.

In addition to HACCP, a tool for the sector itself, the feed sector is currently regulated by a number of Directives aiming to communicate information about the products to other chain actors with the broader goal of improving feed safety. In particular the following Directives are currently in force: Council Directive 79/373/EC on the marketing of compound feeding stuffs; Commission Directive 80/511/EEC authorising, in certain cases, the marketing of compound feeding stuffs in unsealed packages or containers; Commission Directive 84/475/EEC laying down the categories of feed materials which may be used for the purposes of labelling compound feeding stuffs for pet animals; Commission Directive 86/174/EEC fixing the method of calculation for the energy value of compound poultry feed; Commission Regulation (EC) No 223/2003 on labelling requirements relating to the organic production methods for feeding stuffs, compound feeding stuffs and feed materials and amending Council Regulation (EEC) No 2092/91. In addition to traceability requirements set by the EU General Food Law the feed industry is also required to provide information about the exact types and amounts of ingredients used in the form of a label (open label declaration[3]).

The European compound feed industry association is also trying to project a better image of the sector to consumers through the organisation of "open days".[4] In an "open day" the public can see an exhibition of feed production and ask questions about safety and other procedures. The organisation of open days by the feed industry is a worthwhile initiative as it gives the public the opportunity to familiarize itself with the production of feed and confront the producers with questions. Nevertheless, sceptics might argue that at the moment it is primarily a matter of publicity (since the media are also invited) and a face-lift rather than transparency.

Farm, Slaughter, and Processing Sectors

In the farm, slaughter and processing sectors the scope of sustainability related information broadens to cover aspects of animal health and safety and animal welfare and in some cases environmental consequences as well. However, while the scope of information broadens, as this passage will show it covers only a niche of

---

[3]It reformed and strengthened the Council Directive 79/373/EC on the marketing of compound foodstuffs.

[4]The feed day was launched for the first time in June 2005 on the occasion of FEFAC's Annual General Meeting in Brussels.

products. With respect to animal health and safety, a number of rules exist covering areas such as animal nutrition (including medicated foodstuffs), feed and food hygiene, zoonosis, animal by-products, residues and contaminants, control and eradication of animal diseases with a public health impact, pesticides, feed and food additives, vitamins, mineral salts, trace elements and other additives, materials in contact with food, quality and compositional requirements, drinking water and controls. Moreover, a number of directives exist that aim to facilitate the tracking and tracing of animals and thus cover aspects of identification and registration of animals, the identification and registration of animal holdings, tracking and tracing of movements of animals as well as veterinary certificates when entering the EU, and random checks when they are moving within the EU (see Kalfagianni 2006). A number of directives cover the exchange of information between national administrators for the notification in case of diseases, as well as the establishment of a network linking veterinary authorities in different member states (called Amino), since there are no more veterinary checks at the community's internal borders.

There are specific rules covering the welfare of animals and specifically of animals kept in intensive farming (Council Directive 98/58/EC concerns the protection of animals kept for farming purposes). Fundamental for the purposes of this paper, there is currently no EU requirement to communicate information concerning either compliance with basic EU animal welfare rules or more stringent rules and standards to final consumers. Compliance with animal welfare standards at the EU level are only indirectly communicated through the EU label for organic food products,[5] which also communicates information about environmental performance. Likewise the EU organic label is the only communicative tool for environmental performance in the pork sector. This concerns information about a very small amount of organically farmed food within the EU representing only 3.7% of the total EU agricultural area (EEC 2092/91 revision).

Retail Sector[6]

The retail sector has also initiated its own information sharing schemes through its EU association Eurocommerce, with the emphasis resting on food safety. These schemes cover the pork chain but are not confined to it. In particular Eurocommerce has initiated a project, financed by DG-Research and involving a number of EU universities, called FOODTRACE seeking to promote a European

[5]The EU label for organic products was established in 2000 (EC regulation 331/2000), but until now the interest in the logo has been minimal. Organic products continue to be marketed under national or private labels (EEC 2092/91 revision). The Commission believes that an EU logo would help increase sales because it will increase recognition among the EU consumers. For that reason the Commission is currently performing research to investigate tools capable of facilitating the adoption of the EU logo by member states.

[6]HACCP also applies here.

concerted action to develop a *traceability* framework for the whole food chain. The initiative aspires to create a practical framework that can be used by all the actors in the chain, including the international level, to ensure traceability throughout all stages of the chain. The proposed identification scheme is technology independent but technologically supported, enabling its use in developing countries.

In developing FOODTRACE, Eurocommerce draws on the rules and procedures of existing traceability initiatives at the national and international levels. More specifically these include the Traceability of Fish Application of EAN-UCC[7] Standards (EAN International), Traceability of Beef Guidelines (EMEG),[8] Fresh Produce Traceability Guidelines (EAN International), Traceability in the Supply Chain (GENCOD[9] EAN France), and Traceability Implementation (EAN-UCC) project.

Next to FOODTRACE, other European and international initiatives for which Eurocommerce is not responsible are GLOBAL-GAP and the Global Food Safety Initiative. The Global Food Safety Initiative, initiated in 2000 by a group of international retailers, aims to ensure consumer protection, strengthen consumer confidence, set requirements for food safety, and improve cost efficiency throughout the food chain. It has 52 members, currently representing 65% of worldwide food retail revenue. Apart from retailers, global manufacturers such as Unilever and Carrefour also participate in the initiative.

GLOBAL-GAP[10] was initiated in 1997 by a group of retailers belonging to the Euro-retailers Produce Working Group (EUREP). It evolved into an equal partnership of agricultural producers and their retail customers. Initially, it covered only fruits and vegetables, but has been expanded to cover meat products and fish from aquaculture. EUREP-GAP certification is contingent upon completion and verification of a checklist that consists of 254 questions, 41 of which are considered "major musts" and 122 of which are considered "minor musts". Another 91 questions are "shoulds" which are recommended but not required practice. For the pork sector all "major musts" concern traceability and food safety, while information on environmental consequences and animal welfare fall into the categories of "minor musts" or "shoulds". As such the protection of human health and safety remains the central goal of the initiative, while the environment and animal health and welfare remain secondary aspects.

---

[7]EAN (European Article Numbering) is a standard numbering system for Europe that is used by businesses who spread their activities on a global level. It started in 1974 when European manufacturers and distributors from 12 European countries decided to promote an identification system for their products. The actual EAN was formed in 1977 as a non-profit organisation operating under Belgian law with headquarters in Brussels. Due to its increasingly global status it was renamed in EAN International in 1992. UCC (Uniform Code Council) is the North American counterpart. The management of EAN.UCC is operated by GS1, which is a voluntary standards organisation. Today GS1 has 101 members operating in 103 countries. Over one million businesses use EAN-UCC standards.

[8]EMEG is the European Meat Expert Group established within EAN International to develop guidelines for the traceability of beef.

[9]GENCOD (Groupement d'Etudes, de Normalisation et de Codification) is the French branch organisation for EAN.

[10]Until 2001 GLOBAL-GAP has been known as EUREP-GAP.

In sum, currently policies and initiatives for information sharing in the pork sector at the EU level are strongly safety oriented. The memory of BSE as well as the regular appearance of other food scares makes the detection of safety problems the primary concern among the food chain actors and politicians alike. Even though this is important and provides a high degree of transparency in the vertical dimension, the horizontal scope remains narrow. As such, broader sustainability concerns remain in the background.

Transparency in the Pork Production and Consumption System: Initiatives at the Dutch Feed Sector Level

At the Dutch feed sector level, policies and initiatives for information sharing also exist and are frequently stronger than those at the EU level. In addition to the EU regulation on traceability requirements mentioned above targeting the entire chain, the Dutch initiatives were primarily developed on a sectoral, link-by-link, basis.

Feed Sector

As earlier mentioned, many food scandals are associated with the composition or contamination of animal feed and the Dutch feed sector is no exception. More specifically, the feed sector in the Netherlands has been involved in a number of feed scandals in the past (dioxin and sewage waste in feed in 1999) and the present (dioxin in 2004, and more recently dioxin in feed 02/02/2005).[11] To control for feed related scandals the Dutch animal feed industry applies two systems of traceability both of which have been initiated by the Product Board for Animal Feed (PDV).[12] The first is the "Early Warning System" (EWS) and the second the "Tracking and Tracing System". The purpose of the Early Warning System is to identify and eliminate any potential hazards for people and/or animals which may arise despite preventive quality assurance.[13] The Tracking and Tracing System is an integral part of the quality assurance system known as GMP+.[14] The purpose of the Tracking and

[11]The last incident was discovered early on by the competent authorities and was not forwarded in the food chain. However it resulted in negative media attention and the criminal investigation of certain feed companies. Many actors voice the fears that feed companies in the Netherlands circumvent the legal requirements and have false certification labels.

[12]PDV stands *for Productschap Dierenvooder* (Product Board for Animal Feed).

[13]Quality assurance is made through the system of HACCP (Hazard Analysis Critical Control Points). This system is designed to manage all public health hazards involved in food production. HACCP focuses on hygiene and product safety.

[14]GMP stands for Good Manufacturing Practice. HACCP and GMP together make up the GMP+.

Tracing System, as explained by the Product Board (PDV), is to track down irregularities in consignments of animal feed and foodstuffs as quickly and as accurately as possible. Apart from tracking and tracing of contaminated feed the GMP system requires that producers only use feeds having a risk assessment registered in the Feed Risk Assessment Database of the Product Board for Animal Feed. Independent certification bodies issue feed companies a certificate accepted by the Product Board to audit and certify companies in the animal feed industry. Companies contravening the rules risk losing their GMP+ certificate. In the Netherlands around 90% of the feed companies are GMP+ certified.

Farm, Slaughter and Processing Sectors

At the farm, slaughter, and processing sectors a system called IKB (*Integrale Keten Beheersing* or Integrated Chain Management) is the tool through which actors exchange information on products and processes. IKB was introduced in 1992 with the aim of identifying the origin of animals. Currently it has expanded to include other types of information including requirements for information provision on hygiene, animal feed, the use of prohibited growth substances, drugs,[15] and to some extent animal welfare. In the Netherlands approximately 80% of pig farmers participate in the IKB.

While health related issues are communicated exclusively within the chain, issues that concern animal welfare are also communicated to the consumers through labeling. In particular the sector has developed three labels screening different levels of animal welfare. Two of these labels are part of the IKB system (PVE-IKB and PVE-IKB *scharrelvarkens*), and one is related to organic production (EKO). The PVE-IKB label signals conformance to the basic national and EU regulations concerning pig welfare.[16] Therefore it signals compliance. The animal welfare organization in the Netherlands criticizes those standards, however, as being low and lacking sensitivity to the natural behavior of pigs. According to the Dutch Animal Welfare Organization (Dierenbescherming) the main animal welfare problems in the pork sector concern the keeping of pigs in very small places, the construction of the floor, the keeping of sows individually, castrating, tail cutting and corner

[15]The feed industry and veterinarians cooperating with IKB need to comply with either the code of Good Manufacturing Practice (GMP) or the code of Good Veterinarian Practice (GVP). Checks on animal health rules are carried out by the SKV (Foundation of Quality Guarantee of the Veal and Meat Sector) and CBS (Central Bureau of Slaughter Stock Services).

[16]Animal welfare in the Netherlands is covered by Community and national legislation. At the national level the Pig Farming Decree (*Varkensbesluit*) (in force since 1996, amended in 1998) was introduced to implement the Council Directive 91/630/EEC (now replaced by Council Directive 2001/88/EC). The Directive sets the minimum standards for pig welfare at a farm. The Animal Health and Welfare Act (*Gezondheids en welweijnswet voor dieren*) in 1992, (revised in 1994 and 1997), sets minimum standards for the layout, dimensions and hygiene of housing, as well as transport of animals and slaughtering methods.

teeth lowering or extracting, intensity of light, inability to walk outside, use of hormones and preventive medication, and nurturing time.

The other two labels signal higher standards for the welfare of pigs. For example the PVE-IKB *scharrelvarkens* label indicates that pigs are kept in larger rooms better equipped for their comfort and allowed to walk outside. In addition, this prohibits preventive medication and use of growth hormones while allowing other practices such as castration. The EKO label represents the highest standards for animal welfare and also indicates organic production. In addition to the farm level, animal welfare rules apply at the transportation stage and for the slaughtering of pigs. In this respect the first two animal welfare labels comply with standard rules forming part of the IKB system while the EKO label involves higher standards.

In regards to information strategies about the environmental consequences of farming, slaughter, and processing processes, information is currently limited to the farm level. Environmental information at the farm level is also limited to tracking manure and ammonia emissions, of which the pig farming has a large share. Pig production has a large share (30%) of ammonia emissions).[17] In 1998 a minerals accounting system was introduced (MINAS) in order to make farms accountable for their methods, account for differences within sectors, and to stimulate technological development and enterprise. This system, developed jointly by the industry and the government, involves a registration of farm mineral inputs (nitrogen and phosphate from fertilizers and animal feeds), as well as mineral outputs in the form of products and manure. The difference between inputs and outputs, the mineral loss, ends up in the environment. If minerals losses (the difference between input and output of minerals) exceed certain standards then levies are applied. The levy rates are progressive in that the more the standard is exceeded the more the farmer has to pay. In the first years of the application, farmers using MINAS were rewarded by labeling their products as environmentally friendly (*milieukeur* label). However as from 2000, the *milieukeur* label was changed to introduce other environmental themes (in particular energy use for the pork sector, applicable from 2005).

According to some observers *milieukeur* certification standards are not particularly high. For example the losses of biodiversity, groundwater drainage issues, and ecosystem damage, all represent big problems from pig farming activities and are activities that are not covered by the label. In contrast, the EKO label used to indicate biological production covers a wider range of environmental themes associated with pig farming activities. The EKO label indicates that activities beyond the farm level have been carried out in an environmentally friendly way.

This discussion shows that a number of initiatives have been undertaken with respect to information provision on animal welfare and environmental consequences in the farm sector. However, in total the number of pigs slaughtered with

---

[17]According to a policy document on manure and ammonia by LNV (2002) this figure could be reduced by low emission application. The development of low-emission pig units is a recent development and is rather expensive; however costs are expected to fall. These systems are also thought to improve animals' living conditions as well as their health.

any of the above-mentioned labels (except for the standard IKB) was only 111,000 animals. This number represents just 0.6% of the total number of pigs slaughtered in or exported from the Netherlands in 2003–2004 (LTO, *Varkenshouderij in beweging, Matschappelijk verslag varkenshouderij* 2003–2004). As such, the initiatives have not been effective in attracting the participation of a large number of farmers, so far. Moreover, environmental initiatives in particular, by-pass the slaughter and processing sectors whose sustainability impacts are particularly high.

Retail Sector

In the Netherlands, the Dutch retailers' association (CBL) as well as four food retailers (Albert Heijn, Laurus, Superunie and AMS) are involved (the other 28 CBL members do not participate individually) in the Global Food Safety Initiative. The Dutch retailers' organization (CBL) as well as a leading retailer (Albert Heijn) and the Dutch farmers' organization (LTO) participate in the GLOBAL-GAP initiative.

Non-Governmental Organisations (NGOs)

Efforts to improve distribution of information on food (and pork) products and processes have also been initiated by the Dutch consumer organization Consumentenbond. In particular, Consumentenbond developed a proposal for a national regulation (Wok).[18] The proposal was made in parallel with the adoption of the EU Regulation and was stimulated by the Dutch Social and Economic Council's (SER) advice on Sustainable Consumption to the Dutch government. Basically, Wok intended to upgrade the role of consumers to stakeholders, facilitate the work of consumer and other societal organizations, and emphasize the need for government intervention. For this reason, it proposed that food chain actors should provide any relevant information on products and processes demanded by societal actors. In addition they should provide information actively by means of yearly reports or the publishing of information on their websites. The proposal was rejected since both the government and the vast majority of chain actors did not want to adopt a mandatory regulation in this context. It was decided that business should become more "transparent" through the publication of their website and telephone addresses, so that consumers could contact them, if they had specific questions. In addition permission was given to Consumentenbond and other societal organizations to conduct relevant research and publish the results. Finally, it was agreed that food companies should publish yearly reports. This output clearly shows that the ambitious objectives of Consumentenbond for provisioning and

[18]The proposal is named "Weet wat je koopt (Wok)", which means "know what you are buying".

distributing a broad range of sustainability related information in the chain and towards the consumers were seriously undermined in the negotiation process.

In sum, at the Dutch level the horizontal scope of sustainability related information shared among the food chain actors broadens in relation to the EU. The problem remains however, that this information covers only a small range of products thus undermining its potential impact on promoting the sustainability objectives set by the EU and member-states for agricultural and food policies.

Remaining Knowledge to Action Gaps and Concluding Remarks

This paper presented sustainability related information schemes in the pork chain at the Dutch and EU levels. The paper argued that the promotion of the new objectives set for agriculture and food, however, can only be successful if policies for the establishment of transparency in the food chain are promoted simultaneously. In examining the status quo with respect to sustainability related information schemes in the pork chain at the Dutch and EU levels, the paper found that these are limited in scope. The emphasis on safety is disproportionate to other aspects of sustainability, in particular animal welfare and environmental consequences.

Why is this so? Stakeholder interviews reveal a number of reasons why actors in the food chain are reluctant to improve the horizontal scope of transparency so far. More specifically, the costs involved in the production and dissemination of non-safety information is considered a fundamental drawback. Certification and monitoring processes are extremely expensive, and actors voice concerns about the disadvantaged position such costs could place them on, in the face of global competition. Moreover, uncertainty about the willingness of consumers to incur the costs of sustainable information is an additional concern. Finally, some interviewees mentioned that the subjective character of sustainability, that is the diverse interpretation of sustainability concerns in different economic and cultural contexts, is a further obstacle in improving the horizontal dimension of transparency. Such actors are usually multinational corporations operating in a multitude of countries across the globe, and the development of (high) EU standards on sustainability related information they complain could put economic strain on their activities abroad.

These are, of course, legitimate concerns. It would be naïve to suggest that economic considerations of food chain actors should be completely overlooked in favor of environmental and animal welfare concerns. Moreover, some of the economic concerns also relate to the well-being of European farmers whose position today is a far cry from the privileges they enjoyed some years ago. Yet, a number of examples show that even very poor countries are able to meet stringent requirements when they have to (Cook and Iliopoulos 2000; Fulton 2001). The answer, especially at the farm level, seems to be the creation of cooperatives to share the costs of implementation. Thus, Henson et al. (2005) discuss the merits of such an option presenting the example of Hortico Fresh Produce Ltd., in Zimbabwe. Bain and Busch (2004) also refer to the successful formation of two processing cooperatives in

Michigan to meet the challenge of increasingly rigid sustainability standards. Similar efforts could be undertaken in Europe.

In addition, stakeholder interviews reveal that *trust*, or more specifically, the absence of trust, is a serious obstacle to the improvement of the horizontal scope of transparency in the production and consumption system. Trust is usually distinguished in two forms: one based on the other party's intentions and the other based on the other party's abilities (Dooley and Fryxell 1999). Trust in someone's abilities determines whether or not that person or organization will be assigned a certain task. On the other hand, trust in someone's intentions is mostly associated with the risk of facing opportunistic behavior (Bradach and Eccles 1989; Gambetta 1988). It is trust in other actors' intentions that is currently limited in the cases under study. The low levels of trust can be explained in terms of the food scandals that have shaken the food chain, but also because of the conflicting roles actors involved in the sustainability debate often assume. Thus, business actors complain that civil society organizations are never satisfied with their efforts; instead as soon as a specific target is achieved they always set the next, usually more stringent, one. In a similar vein, civil society organizations protest that business actors often act strategically and try to avoid their responsibilities; moreover, they feel that (powerful) business actors have better access to decision-making processes than they do.

What can be done to improve the current situation? First of all, public actors can create the conditions necessary for trust improvement, by providing equal access to both business and civil society actors on decision-making procedures concerning transparency. Currently, even though access is provided, civil society assumes a consultative role and therefore, has limited opportunities to influence actual decisions before they have been made. In addition, actors themselves have to make steps towards improving trust relationships. Private–Private Partnerships in the monitoring and implementation of schemes related to transparency could work towards that direction. This suggestion also invites a note of caution, however, as the involvement of civil society in private initiatives for sustainable information, although desirable, could merely act as a legitimation factor.

It is, perhaps, too soon to expect dramatic transformations in the information and communication processes for sustainability in the food chain. Yet, efforts toward that direction from all the societal and political actors involved are necessary (even though not sufficient) conditions for the improvement of sustainability in the food system as a whole.

Appendix

Table A.1 Overview of sector-oriented policies and initiatives for sustainability related transparency in the pork chain at the EU and Dutch levels

	EU	The Netherlands
Feed sector	HACCP	HACCP
	European Feed Manufacturing Code	Early Warning System
	International Feed Ingredient Standard	Tracking and Tracing System
	Open Feed Days	
Farm, slaughter and processing sectors	Amino	IKB
	Animal Welfare Requirements	PVE-IKB scharrelvarkens
	EU Organic Label	Millieukeur
		EKO label
Retail sector	FOODTRACE	FOODTRACE
	GLOBAL-GAP	GLOBAL-GAP
	Global Food Safety Initiative	Global Food Safety Initiative
NGOs		*Weet wat je koop* (know what you are buying) initiative

References

Andreasen C, Strynth H, Streibig JC (1996) Decline of the flora in Danish arable fields. J Appl Ecol 33:619–626

Bain C, Busch L (2004) Standards and Strategies in the Michigan Blueberry Industry. Michigan Agricultural Experiment Station Report 585, Michigan State University, East Lasing, MI

Baldock D (1990) Agriculture and habitat loss in Europe. CAP Discussion Paper. Gland, Switzerland

Biesiot W, Moll HC (1995) Reduction of CO2 emissions by lifestyle changes. NRP Program Office, Bilthoven

Bowler IR (1985) Agriculture under the Common Agricultural Policy: a geography. Manchester University Press, Manchester

Bradach JL, Eccles RG (1989) Price, authority and trust: from ideal types to plural forms. Annu Rev Sociol 15:97–118

Buzby JC (2002) The graying of America: older adults at risk from complications from microbial foodborne illness. Food Rev 25(2):30–35

Cook ML, Iliopoulos C (2000) Ill-defined property rights in collective action: the case of US agricultural cooperatives. In: Menard C (ed) Institutions, contracts and organisations: perspectives from new institutional economics. Edward Elgar, Northampton, pp 335–348

Delbeare B, Drucker G, Lina P, Rientjes S, Vinken H, Wascer D (1998) Facts and figures on Europe's biodiversity. State and Trends 1998–1999. ECNC, The Netherlands

Dooley RS, Fryxell GE (1999) Attaining decision quality and commitment from dissent: the moderating effects of loyalty and competence in strategic decision-making teams. Acad Manag J 42(4):389–402

Fuller RJ, Gregory RD, Gibbons DW, Marchant JH, Wilson JD, Baille SR, Carter N (1995) Population declines and range contractions among lowland farmland birds in Britain. Conserv Biol 9(6):1425–1441

Fulton M (2001) Traditional versus new generation cooperatives. In: Merrett CD, Walzer N (eds) A cooperative approach to local economic development. Quorum Books, Westport, CT, pp 11–24
Gambetta D (1988) Can we trust trust? In: Gambetta D (ed) Making and braking cooperative relations. Basil Blackwell, Oxford, pp 213–237
Gardner B (1996) European Agriculture: Policies, Production and Trade. Routledge, London
Goodman D, Redclift M (1991) Refashioning nature: food, ecology and culture. Routledge, London
Henson S, Masakure O, Boselie D (2005) Private food safety and quality standards for fresh produce exporters: the case of hortico agrisystems, Zimbabwe. Food Policy 30(4):371–384
Hill MJ (1999) Meat and colo-rectal cancer. Proc Nutr Soc 58(2):261–264
Hoekstra J (2002) Officium nobile, officium durum. Universiteit van Amsterdam, Amsterdam
Kalfagianni A (2006) Transparency in the food chain. Policies and politics. Twente University Press, The Netherlands
Kramer KJ, Moll HC, Nonhebel S (1999) Total greenhouse gas emissions related to the Dutch crop production system. Agric Ecosyst Environ 72:9–16
Lijinski W (1999) Research papers N-nitroso compounds in the diet. Mutat Res 443(1–2):129–138
Manhoudt AGE, de Snoo GR (2003) A quantitative survey of semi-natural habitats on Dutch arable farms. Agric Ecosyst Environ 97:235–240
Marsden T, Munton R, Ward N, Whatmore S (1996) Agricultural geography and the political economy approach: a review. Econ Geogr 72:361–75
McKnight GM, Duncan CW, Leifert C, Golden MH (1999) Dietary nitrate in man: friend or foe? Br J Nutr 81:349–358
McMichael P (1997) Rethinking globalization: the agrarian question revisited. Rev Int Political Econ 4:630–662
Navarro A, Diaz MP, Lantieri Muñoz SE, MJ EAR (2003) Characterisation of meat consumption and risk of collateral cancer in Cordoba, Argentina. Nutrition 19(1):7–10
Norat T, Lukanova A, Ferrari P, Riboli E (2002) Meat consumption and collateral cancer risk: dose response meta-analysis of epidemiological studies. Int J Cancer 98(2):241–256
Peters PK, Pike MC, Garabrandt D (1992) Diet and colon cancer in Los Angeles County. Cancer Causes Control 3:457–463
Smith D, Riethmuller P (2000) Consumer concerns about food safety in Australia and Japan. Br Food J 102(11):838–555
Tansey G, Worsley T (1995) The food system: a guide. Earthscan, London
Tauxe RV (2002) Emerging foodborne pathogens. Int J Food Microbiol 78(1–2):31–41
UNEP (2000) Cleaner production assessment in meat processing. Industrial Sector Guide, United Nations Environmental Programme (UNEP), Nairobi, Kenya
Van Ek, Remco Witte, Jan-Phillip M. Runhaar, Han Klijn (2000) Ecological effects of water management in the Netherlands: the model DEMNAT. In: Ecological Engineering, 16(1):127–141
Van der Poel WHM, Verschoor F, van der Heide L, Herrera MI (2001) Hepatitis E virus sequences in swine related to sequences in humans: The Netherlands. Emerg Infect Dis 7:970–976
Van Leeuwen G (2003) Farm retirement in the Netherlands: 1950–2002. In: LEI report 6.03.07, Policies for Agriculture in Poland and the Netherlands, LEI
Ward N, Almas R (1997) Explaining change in the international agro-food system. Rev Int Political Econ 4:611–629
Wilkens LR, Kadir MM, Kolonel LN (1996) Risk factor for lower urinary tract cancer: the role of total fluid consumption, nitrates and nitrosamines, and selected foods. Cancer Epidemiol 5:161–166
Willet WC, Stampfer MJ, Colditz GA (1990) Relation of fat meat and fiber intake to the risk of colon cancer in a prospective study among women. N Engl J Med 323:1664–1670
Yeung RMW, Morris J (2001) Food safety risk: consumer perception and purchase behaviour. Br Food J 103(3):170–186

Chapter 10
Sustainable Consumption by Certification: The Case of Coffee

Arnold Tukker

Rationale

The goals to be pursued under a sustainability and sustainable consumption and production agenda often are framed and interpreted in very different ways by different actors. Fischer Kowalski et al. (1994) speak in this context of a 'plethora of paradigms'; there have been ample debates on 'strong' and 'weak' sustainability (Munasinghe and Shearer 1995), and the role of the free market has been described as both supportive as opposing sustainable development (Scherhorn 2005; Prahalad 2004).

Yet, a few elements seem commonly accepted (UN 2002; see also the review in Tukker 2008). With regard to the environmental dimension, radical improvements of resource-efficiency and reduction in environmental impacts are to be strived for. With regard to the economic dimension, the goal is to realise equitable growth – which particularly in developing countries implies a radical enhancement of wealth per capita. And in relation, with regard to the social dimension, in order to realise acceptable social (labor and other) circumstances, particularly again in developing countries radical improvements are needed.

The production consumption system of coffee, central to this chapter, is related to all these elements. Food PCS in general are responsible for a main part of the life cycle impacts of consumption (Tukker et al. 2006; Hertwich 2005). The coffee chain crosses the North–South divide, with relatively poor coffee farmers at the production side and relatively powerful players like coffee roasters and retailers in the North. As will be discussed in more detail later, liberation of the market in the 1990s caused a 'race to the bottom' at the production side, threatening basic minimum social and environmental standards. Coffee is – after oil – one of the most important commodities traded world-wide, and if solutions can be developed for making the coffee PCS more sustainable, this can serve as a role model for other commodities that are important in the North–South relation. Furthermore, the case links consumption with

A. Tukker (✉)
Program Manager Sustainable Innovation, TNO, Delft, Netherlands
e-mail: arnold.tukker@tno.nl

L. Lebel et al. (eds.), *Sustainable Production Consumption Systems: Knowledge, Engagement and Practice*, DOI 10.1007/978-90-481-3090-0_10,

production: as will be explained later awareness of players at the consumption side, formed an important driver for action to solve problems at the production side.

The main pathway pursued to solve sustainability problems in the coffee chain are certification mechanisms providing minimum demands with regard to social and environmental standards at the production side.[1] This, then, also directly points at some limitations in the case. First, where consumers do play a role in the change process, the real changes take place at the production and product side. Behavioral changes hardly have to take place. Second, the case does not imply a radical overhaul of technologies, practices, and relations in the system. It is hence not an example of radical or systemic change.

In this paper we will discuss first some theoretical deliberations that can help to understand changes towards sustainability in production–consumption systems. We then will analyse the case proper: the actors, their actions, and the ultimate result. In a reflective chapter, we will discuss what knowledge actors have used, how they linked knowledge to action, and what leverage can be created to improve the change to sustainability in the case at stake.

Changes to Sustainable Consumption and Production: Conceptual Deliberations

The change to sustainable production consumption systems in an area like the coffee chain is a systemic challenge where networks of actors along the production–consumption chain have to change behaviour in a context shaped by institutions and framework conditions (compare Fig. 10.1). We will present two strands of theory that helps to understand such change, and that can help to understand the potential and limitations of certification schemes in the coffee chain.

Porter's Five Forces Model

Around 1980, Michael Porter (1985) introduced his highly influential concepts of the five forces model and the value chain (which correspond de facto what in this book is called production and consumption systems – see Fig. 10.2). According to Porter, the attractiveness of an industry is defined by five external forces:

1. Competition between present firms
2. Threat of new entrants
3. Bargaining power of suppliers

[1] Other examples can be found in forestry (Forest Stewardship Council – FSC), fishery (Marine Stewardship Council – MSC) and agriculture (see e.g. Belz 2004) who showed how certification of organic or semi-organic agricultural products in Switzerland – after other pressures came in place – formed a stepping stone for a full transition of Swiss agriculture to (semi-)organic practices.

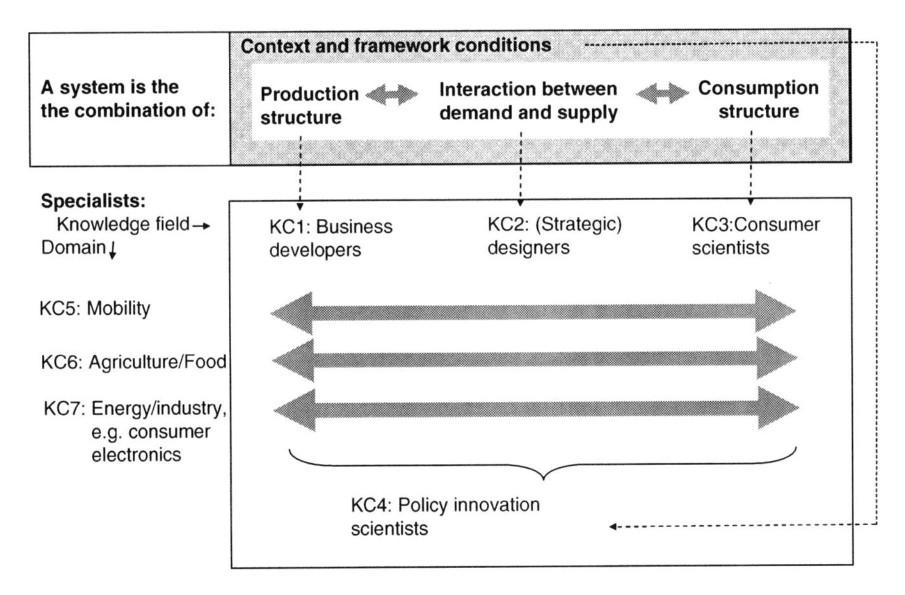

Fig. 10.1 The systemic nature of the SCP challenge

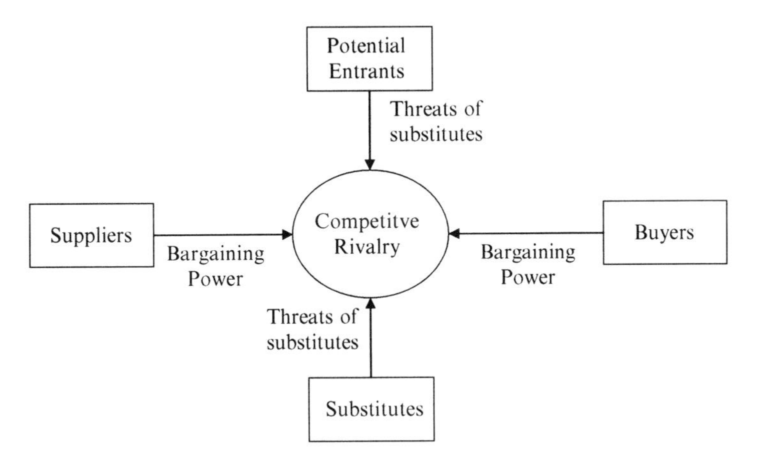

Fig. 10.2 Porter's (1985) five forces model

4. Bargaining power of consumers
5. Threat of substitute products

A company can flourish if it is able to fence off new entrants and substitutes, and if its bargaining power vis a vis customers/distributors and suppliers is high. Threats of new entry depends for instance on the extent to which entry barriers are present in an industry (e.g. economies of scale, capital requirements, access to distribution channels and so on). The links (and actors) in a production–consumption value chain where the five forces turn out to be most favorable tend to be the places where uniqueness and the power in the value chain is concentrated. In sum, these are the places where most of the value created in the chain can be captured, but also where the power can be found to change practices in the chain (compare Wise and Baumgartner 1999; Davies et al. 2003; Hartford 2005).

The Multi-level and Multi-phase Model

Where Porter's value chain concept helps to map a production- and consumption chain, in the late 1990s more elaborate system concepts were proposed that contribute to a broader understanding of radical change in such systems. Briefly said, authors such as Rip and Kemp (1998), Geels (2002), and Berkhout et al. (2004) developed a system concept of three levels, in which innovation and change has to be understood as an evolutionary processes of variation and selection (see Fig. 10.3). The meso- level is called the 'technological paradigm' or 'socio-technical regime'. It depicts the mainstream way of doing things as an interconnected set of elements like technologies, knowledge, markets, infrastructure, cultural values, practices and symbolic meanings (Nelson and Winter 1982; Dosi 1982; Geels 2002; see Fig. 10.4). The micro-level consists of socio-technical niches: relatively limited areas, in which new socio-technical systems and practices can be developed and tested under rather protected circumstances (e.g. a market niche where protection is provided by a specific market demand). Finally, there is a macro-level: the socio-technical landscape. This is a set of fairly stable factors that are in principle external to (actors in) the regime. Examples include the existing material infrastructure, geopolitical realities like location of natural resources, but also the existing culture, life-style, demography, or trends therein, etc.

We now can understand why radical and systemic innovations often are so difficult to realise. The interconnected elements in the regime cannot be changed

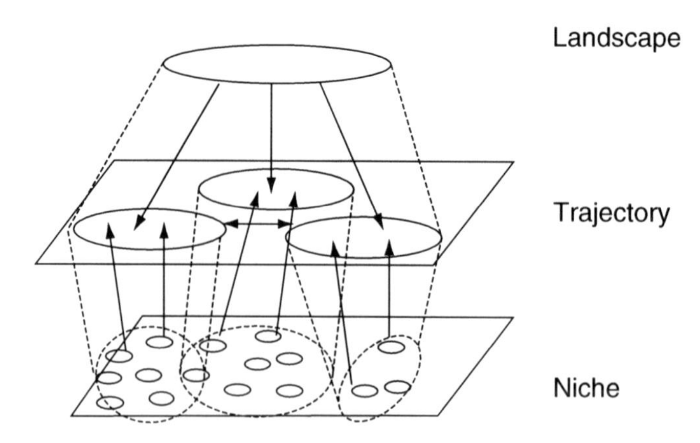

Fig. 10.3 A multi-level perspective on innovation

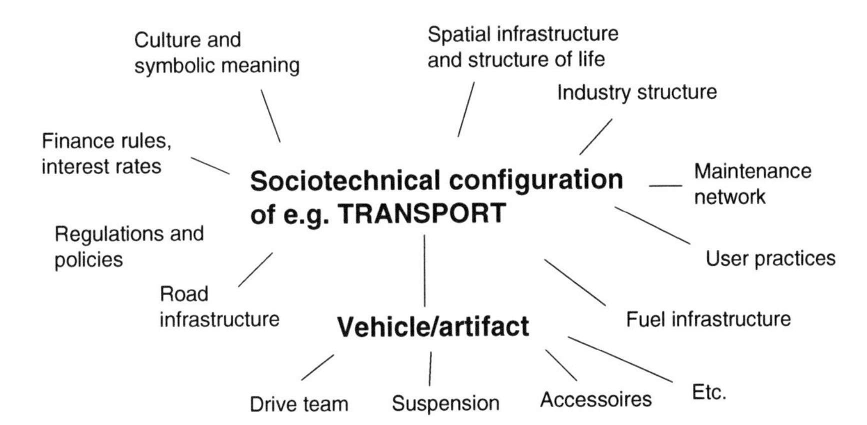

Fig. 10.4 An example of a socio-technical regime: Mobility (Geels 2002)

independently, but must co-evolve.[2] Landscape factors channels the development of regimes. Such systemic forces simply surpass the power of individual actors, and provide a powerful explanation why knowledge-action gaps exist: though individual actors may be aware of desirability for change, given their place in the system they lack the power, or can't be expected to have the interest, to pursue change.[3]

Case Study: Certification in the Production Consumption System of Coffee

Introduction and Hypotheses

The theories in Chapter 2 suggest some important limitations with regard to certification in the coffee PCS. The resistance to change of the dominant regime suggests that a novel certification system may see two archetypical implementation routes. Certification systems that pose radically new demands may easily be

[2] For instance: implementing a new mobility system depending less on petrol using cars probably needs a change of cultural values (the consumer preference for car ownership), change in infrastructure, and development of different fuel provision systems like hydrogen or electricity tap points.

[3] The concept of 'transition management' has been proposed to stimulate change in such 'locked in' situations (Kemp and Rotmans 2004; Kemp et al. 2006). It is basically a reflective governance process that tries to articulate the systemic problem with a group of front runner actors sensitive to the need for change, develop a long term vision, and to identify short term actions and experiments that help to make the system more fluid. Change is obviously highly stimulated when 'events' change the landscape (e.g. discovery of a gas field that sparked the move away from coal as energy carrier in the Netherlands (Correljé and Verbong 2004), internal weaknesses and contradictions develop in the regime, and promising alternative practices are available in niches (cf Te Riele et al. 2000).

rejected by the dominant regime players, and hence be confined to niche markets; and certification systems that after a normal diffusion process make it to the 'mainstream' must somehow fit with the interest of the main players in the regime. We will now analyse what certification schemes are operational in the coffee market, and what impact they have had, after introducing (regime-)actors in the chain and the context (landscape) in which they operate.

Actors in the Coffee Chain

The coffee chain consists of the following steps (e.g. Slob and Oldenziel 2003; Consumers International 2005)[4]:

1. Growing. It takes a coffee plant 3 to 5 years to reach full yield for a period of at least 20 years. Supply is hence price inelastic on short term. Quality is determined by the choice of species, the altitude at which the coffee is grown, and the approach to picking (which determines the amount of unripe berries).
2. Preliminary processing (dry or wet, usually on the farm). The (dominant) dry process implies sorting and cleaning to remove twigs, dirt and damaged/unripe cherries, followed by drying in the sun or via mechanical means. In the wet process fleshy pulp is first removed mechanically followed by a fermentation process, washing, and drying in the sun or machine to give parchment coffee. The right level of drying is crucial for on quality, with fungal attacks and broken beans as trade offs.
3. Secondary processing (hulling). The dry cherry or washed parchment is treated in a hulling process that removes the outer layers of the dried cherry, or the parchment from wet processing. This is followed by cleaning, screening and sorting. This process is too large scale to be owned by small farmers.
4. Export/import. Usually, the coffee then goes to a trading company for export.
5. Roasting. Roasting takes place close to consumers because of the short shelf-life of roasted coffee compared to green beans. A key function of roasters is also blending of coffee types to reach a desired and recognisable taste.
6. Retail and other points of sale.
7. Consumption.

The different steps in the value chain have quite different characteristics As for coffee bean production, over 70% of the coffee is produced by a great number of smallholders of less than 10 ha (Potts 2003:6). But in other steps, the chain has become highly concentrated. Just ten roasters are responsible for 63% of the global sales of processed coffee, while five trading companies account for 40% of the

[4] Of course steps in the chain are not always distinct and vertical integration occurs: roasters may have their own import organisations, or may create their own retail outlets, etc.

imports (Potts 2003:4, note 9). In turn, roasters are in many countries dependent on just 4 or 5 major retail chains for reaching the final consumer.[5]

Developments Between 1985 and 2005

Pre-1989: Regulated Markets

In the period between 1962 and 1989 a rather regulated world market existed for coffee, under the International Coffee Agreement (ICA) regime. The first ICA was signed in 1962 by most coffee producing and consuming countries. Under this system, a target price or price band for coffee was set. Export quotas were allocated to producing countries. Quotas were tightened in periods of low demand and low (indicator) prices; in times of high demand and high prices and quotas were relaxed. In this way, a certain price level was always guaranteed. Furthermore, the system provided a balance of power: in the ICA system, no one actor in the global value network for coffee had a truly dominant position (be it in producing or consuming countries). This regulated market often lead to the situation that in producing countries government tried to exercise control over production volumes and trade, and also over quality standards with regard to coffee for export. Roasters, who had a powerful position in consuming countries, could not exercise much control over their supply chain given the ICA quota system and government interventions in producing countries (Ponte 2002:34–35).[6] For coffee farmers, this system had both positive and negative effects. Positive points were that the coffee prices were at least stable within one season and there were quality incentives; a negative point was that in many producing countries the trade system was bureaucratic, so that only a relatively low fraction of the export price was transferred to producers (Ponte 2002:15–16).

Enter the Free Market

The breakdown of the ICA system and the resulting liberalisation of markets resulted in massive shifts in the power position of actors in the global coffee value network, with a variety of consequences:

---

[5] The market for catering, restaurants and coffee bars is probably a different matter; indeed, roasters like Douwe Egberts/Sara Lee have entered the catering market themselves by offering coffee solutions directly to professional users (Tukker and Tischner 2006).

[6] Although roasters and retailers were then already capable of improving their position vis a vis buyers considerably, the price paid by the consumer in the US rose with 240% between 1975 and 1993, whereas the international market price declined with 18% in the same period (Potts 2003).

1. Initially, the liberation of markets resulted in buyer competition. Farmers were able to demand prompter payment and the fraction of the export price that became available for farmers rose.
2. On the longer run, the lack of control over production volumes led to a market with seller competition and fierce price competition as a consequence (see also next section). With the government (who also had influenced quality control) in producer countries retreating, the fierce price competition also put product quality under severe pressure.[7]
3. The position of the co-operatives weakened vis a vis private traders, who in general could operate with more flexibility and agility.
4. The power in the system clearly shifted towards the nodes in the value network: particularly the roasters and to a lesser extent the traders – becoming the parties who were capable to capture most of the value created in the system.

Enter Vietnam

Apart from the demise of the ICA, one of the most dramatic changes in the international coffee landscape was the expansion of coffee production, particularly in Vietnam. Production grew from a tiny 8,400 tpa in 1980, via 200,000 tpa in 1995, until 900,000 tpa (15 million bags) in 2001. This made Vietnam the second largest producer country after Brazil (CI 2005), with almost 15% of the world market of 122 million bags in 2002 (CI 2005:17). Where in the mid-1990s the global scale supply and demand were still sufficiently in balance to ensure reasonable market prices, after 1997 the entrance of Vietnam – but also expansion of production elsewhere – lead to a great overproduction and a sharp drop in prices as result. Coffee prices fell 70% between 1997 and 2001, reaching the lowest level in 30 years and the lowest level in 100 years when corrected for inflation (CI 2005:16). Since this low in 2001, prices have risen about 30%.

Conclusions

The opening up of markets and the instability of the market in the late 1990s had important consequences for the structure of the coffee value chain, which can be well understood with the Porter model.

With the liberalisation of the market, the initial inflexibility of supply played out in favour of the farmer: in the sellers market they could pull more added value to

[7] A point even aggravated by the fact that it is rather difficult to control coffee quality. Where larger estates produce sufficiently volumes to warrant testing on quality, for the small batches from small farmers testing often is not profitable anymore. This forces small farmers almost always to produce for the low-end market high quality is not recognized (Ponte 2002:16).

them from the export prices, and organisations with less flexibility like co-operatives were by-passed and in many cases went bankrupt. However, the opportunities for free trade were quickly used by new entrants in the coffee farming step to join a then still profitable market. Furthermore, actors in nodes in the value chain grasped the opportunity to strengthen their positions further: particularly among the importers and exporters a strong concentration took place, filling the (power) gap left by ICA and state controlled trade. The net effect was that once around 1997 the sellers market turned into a buyers market due to the supply of new entrants, the small farmers inevitably draw the short straw. The value chain had been restructured in such a way that all the pressure of the price competition could be passed through into their ballpark by the more powerful players in the chain. After all, virtually all developments in the end lowered the bargaining power of farmers: due to a lack of state control and quota systems anyone could enter their market; the exports were now controlled by an oligopoly; and institutions like co-operations that could give small farmers some control over the downstream value chain had been damaged or vanished. Sheer price competition, and hence the production of even lower quality products, that in turn would receive again lower prices, seemed the negative spiral in which the sector was caught.

Certification Schemes: Approach, Market Share, and Impacts

Introduction: Overview of Schemes

To oppose this 'race to the bottom', since the late 1970s a variety of certification schemes was put in place that aims to ensure certain minimum conditions to be met at particularly the production side of the coffee chain. Such schemes have various origins, philosophies and goals. Some were started by NGOs and developing country movements based on the conviction that the richer developed world should not exploit the developing world but pay a fair price for imported commodities thus allowing for reasonable living conditions and wages in exporting countries. The Max Havelaar brand is probably the best-known example of this 'Fair Trade' movement. Other schemes had their roots in environmental concerns, and focused on agricultural production methods with no or limited environmental damage. 'Organic' certification schemes are the most prominent examples of this stream. And finally, with the rise of the importance of brand value and transparency of supply chains, companies at the consumption side of the chain (like roasters, retailers, and other consumer outlet) understood that a certain level of 'corporate social responsibility' was in their interest. Consumers simply expect 'good behaviour' of their suppliers, and a minimum quality level in the supply chain is beneficial for the controllability and continuity of downstream processes. This lead to a third form of what I would call 'market initiated' form of certification, of which the Utz Certified[8] scheme and initiatives of companies like Starbucks are clear examples. In the next sections we will discuss the following schemes (see Table 10.1 for a comparison):

- Fair trade labels
- Organic labels
- Utz certified coffee
- Rainforest Alliance/Sustainable Agriculture Network certified coffee

Fair Trade/Max Havelaar: Target the Consumer by a Dedicated Brand

The fair trade movement has as its main aim to ensure minimum producer prices for goods sold by developing countries to the West. Over time, national fair trade groups organised themselves into the "Fairtrade Labelling Organisation" (FLO). The FLO sets minimum prices per type of coffee and the region. If the market prices are above the minimum price, a fair trade premium of 5 US cts per lb of green coffee applies. Coffee certified as organic receives an additional premium of 15 cts per lb of green coffee.

The fair trade system is geared towards smallholders who have to be organised in producer groups or co-operatives, which must be democratically run and politically independent. The fair trade premium is paid to this group for use in community projects. The fair trade label also demands compliance with some environmental standards, such as a ban on the use of most toxic pesticides and the application of integrated crop management (CI 2005). Furthermore, compliance with standards of the International Labour Organisation (ILO) is also demanded.

NGO pressure on mainstream roaster companies lead to the inclusion of fair trade coffee in some of their specialty brands. After around two decades of existence of the system around 24,000 tpa (or 0.4% of global demand) of coffee is now sold under the fair trade label.

Though this volume is not negligible, also compared to other certification schemes, there is widespread feeling that the FLO system in the end will stay serving a niche market. The price gaps with regular coffee brands tends to be substantial, sometimes 50% or more compared to regular coffee brands,[9] and well beyond the

---

[8] Utz Certified started its work under the name 'Utz Kapeh', literally 'good coffee' in the Mayan Language. Some documentation we refer to still uses the old name that was used before the name change in March 2007. See http://www.utzcertified.org/logoexplanation/logoexplanation_b2c.php, accessed 11 November 2008.

[9] This is in part caused by market logic in the retail sector. Attracting clients is important, and an often used strategy to lure customer into shops is to price some basic, recognisable commodities like coffee very low, giving the shop a cost-effective image. As shown by theories on bounded rationality, shoppers are usually not capable to do an integrated price comparison over the dozens or even hundreds of products they buy, allowing supermarkets to cross-subsidize commodities like coffee via higher premiums on other products. Another issue is that shelf space at supermarkets is increasingly at a premium – indeed, supermarket chains have become such powerful players in the value chain that they now can demand an upfront payment for granting producers the 'right' to have their products displayed on the shelves. This logic, perfectly explainable by the Porter model, obviously is not in favour of the smaller players that tend to produce FLO labelled products.

Table 10.1 Comparison of certification schemes (based on CI 2005; Slob and Oldenziel 2003)

	Fair Trade	Organic	Utz Certified	Rainforest Alliance
Goal	Ensure equity by fair prices for small producers of crops	Ensure organic production of crops	Enable brand owners and coffee producers to claim that coffee has been produced under socially and environmentally responsible standards	Integrate productive agriculture, biodiversity conservation, and human development
Strategy	Promote the fair trade brand to the consumer, and receive a premium price	Promote the organic brand to the consumer, and receive a premium price	Target interests of brand owners (and producers), who have a consumer-induced interest to ensure that 'responsibility' is a natural element of their brand	Focus on mainstream and specialty brands
Demands	Only smallholders organised in democratically run co-operatives without relations with the state can participate. Environmental standard like bans on certain pesticides	Use of synthetic nutrients and crop protection methods banned, soil conservation demands	Good agricultural practice and social demands derived from the EurepGAP and ILO standards.	Relatively stringent environmental and social demands, including restoration of native forest reserves
Factual price premium	Minimum price of US $ 1.01–1.21 per lb depending on price and origin	US$ 0.15–0.30 per lb (2003)	US$ 0.04 per lb (average 2004)	US$ 0.05–0.15 per lb (Brazil)
Market volume around 2004–2005 (world total: 6–7 Mio tpa)	24 ktpa	40 ktpa	30 ktpa	p.m.

'willingness to pay' of mainstream consumers (compare CI 2005:43).[10] Inevitably, in order to explain why such a high price premium is asked, the certification logo has to become a visible part of the coffee brand. Despite this success of the initiative in showing that it is possible to do business on the basis of confidence, quality, long term relations and fair prices, scepticism persists whether the approach can solve the problems in the coffee system. According to Slob and Oldenziel (2003), "The FLO does not aspire to work with medium or large-scale producers .... In view of its limited span of impact, however, one could say that the FLO initiative on coffee is neither the definitive answer to the malaise on the world market, nor to everyday problems of bulk coffee producers".

Organic Certification

Organic certification systems usually require the use of non-chemical nutrient supply and plant protection methods, as well as practices to enhance the soil structure. In quite some cases, organic certification standards are set or regulated by governments. For instance, to avoid internal trade barriers, the EU has developed regulations on organic production.[11] Demands formulated by the International Federation of Organic Movements (IFOAM) are usually taking as initial guidance for developing (e.g. crop-) specific standards. In the case of coffee, a transition period of 3 years is usually required before the product can be sold as organic. Production in 2002/2003 was about 48,000 tons and, given the price premium for organic coffee in terms of value, the most important type of certified coffee.

Utz Certified: Target Producers by Offering Transparency and Credibility

Where many other certification bodies try to target consumers and end up being a specialty brand for niche markets, Utz Certified has a different philosophy. Their vision is to ensure via certification that minimum environmental and social

---

[10]This goes back to the famous 'consumer'–'citizen' dichotomy described by so many authors. In their role as citizen, people of course want to support solving big problems like poverty in the world. As we will explain later, in this role their influence on production and consumption chains is probably more effective than in their role of consumers. Where there are exceptions, across the board consumers are in practice ready to a pay only a limited price premium (say more than 5–10%) for products which display additional characteristics like 'organic', 'ecological', or 'fairtrade'.

[11]Council Regulation (EEC) No 2092/91 of 24 June 1991 on organic production of agricultural products and indications referring thereto on agricultural products and foodstuffs.

requirements are met in the coffee chain – as a world standard. This ambition to become mainstream leaves no option but to work out a strategy that in the end will win the main players in the coffee scheme to this ideal. As adjunct Utz Certified director Lucas Simons puts it: "Leading roasters and retailers combine Utz Certified as a baseline scheme for their mainstream ranges with fair trade or organic programs for special labelled products. This is the future we envisage". Becoming in just 6 years of existence the biggest coffee certifier globally, and a market share of 25% in the country of origin of the Netherlands, it seems that Utz Certified has developed a formula that has at least a chance to make this ambition true.

The Utz Certified history started in the end of the 1990s, when Ahold, the major Dutch retailer and also owner of a coffee roaster company and -brand, teamed up with a group of Guatemalan grower-exporters. They developed a 'Code of conduct' to provide basic assurance for food safety and environmentally and socially appropriate growing practices. The Utz Certified foundation was set up, which independently from Ahold further developed and disseminated the scheme. Currently, the Utz Certified approach consists of the following elements (Utz Certified 2006a, b, Slob and Oldenziel 2003; CI 2005):

1. Coffee producers have to work according to a 'code of conduct', consisting of a list of 'major musts' (100% compliance required), 'minor musts' (95% compliance required) and 'recommendations' (voluntary). The Code of conduct embodies the four core labour standards of the International Labour Organisation (ILO) and environmental standards following the Eurepgap framework for Good Agricultural Practice (GAP) of the Euro-Retailer produce Working Group (Eurep). Unlike fair trade systems, Utz Certified does not require minimum prices.
2. Additionally, parties in the coffee chain must comply with the Utz Certified Chain of Custody Requirements.
3. Parties making on product claims about Utz Certified must be certified by a third party certification body, approved by Utz Certified. For blended coffee products to which the Utz Certified claim can be applied, at least 90% of the content must comply with Utz certification standards.
4. Unlike fair trade standards that focus on small producers Utz Certified directs itself also towards medium and large scale farms. Utz Certified also aims to channel technical assistance to producers, and to improve access to credit facilities.[12]
5. A key objective is that certified producers can differentiate themselves from conventional growers, and can easily link up with responsible buyers. As a facilitation instrument, Utz Certified maintains lists of certified producers and registered buyers.

In sum, the Utz Certified strategy seems very much to play the card of the interest that roasters, retailers, and other brand owners have to ensure that responsibility is

[12] This 'supplementary policy' is of importance for smaller coffee growers, generally seen as a group for which implementation of certification schemes is relatively costly. For instance, with support of the Douwe Egberts Foundation and Ibero Uganda 3,000 small scale farms in the Luwero district in Uganda were able to acquire Utz certification.

a natural element of their brand identity, and the need to control the quality of the supply chain as a consequence of this.[13] Consumers play a surprisingly small role. Their expectation that mainstream brands should embrace responsible production forms the final driver – but since it is seen as a basic attribute of the products they buy, certification is of limited use for product differentiation or to ask (significant) premium prices. Targeting roasters, retailers and coffee producers is hence the main goal of Utz Certified, and their key instruments (ensuring minimum standards for production and transparency in the supply chain) are tailored to this goal. Intervening in price setting – a fundamental philosophical point of discussion, see later – is even though preferred by Utz Certified, not a realistic option. This probably would lead to a loss of a lot of the current support from roasters and retailers. Utz Certified tries to strengthen the (bargaining) position of farmers by publishing weekly the price premiums paid for Utz Certified certified coffee. In practice, this premium is about 4–5 cents per lb.

Rainforest Alliance

The Rainforest Alliance is an international conservation organisation with as mission to protect ecosystems and wildlife and people living in them. Their 'Sustainable Agricultural Network (SAN)' is managed jointly with a network of Latin American NGO partner organisations. SAN have developed guidelines for responsible management of various crops, like bananas, cacao, citrus and coffee. Farms complying with SAN standards may use the Rainforest alliance label when marketing their products. Their scheme currently focuses mainly on Latin America, although also operations in the Philippines have been certified. Demands include compliance with ILO standards, and relatively stringent environmental standards based on integrated pest management.[14] Like in the case of Utz Certified, no intervention on price setting is foreseen, and the scheme has been joined by all types of producers, from smallholder cooperatives to large agribusinesses. SAN rather than independent certifiers carry out the evaluation of firms.

The Rainforest Alliance demands that for blended coffee products to be able to carry their label, at least 30% of the content must be certified. They made a considerable effort to persuade large companies to buy certified coffee, including Kraft and Procter and Gamble. Like Utz Certified, the Rainforest Alliance hence set itself as a goal to advance the availability of certified sustainable coffees in the mainstream market (CI 2005:25). Compared to Utz Certified, the Rainforest Alliance scheme is more costly for producers due to the fact their environmental and social demands are somewhat more stringent, particularly the requirement to restore native ecosystems.

---

[13] In the words of Utz Certified Director David Rosenberg: 'Utz Kapeh' makes it possible to make a credible claim about their sourcing across the board, rather than for a small segment of premium prices products'.

[14] IPM allows for limited use of agro-chemicals with strict controls.

Reflection: Knowledge, Links to Action and Leverage

Introduction

So now back to the core of this paper: given these backgrounds, what can be the contribution of coffee certification schemes to realising sustainability in the coffee production consumption system?. We will explore this question by discussing knowledge about sustainability problems, its transformation into action, and additional points of leverage that could be created.

Knowledge

The case shows that applying a simple 'knowledge-action' model is probably not the best angle to analyse the problem of (un)sustainability in the coffee chain. Such a model implicitly fails to problematize knowledge, and also suggests that it is primarily knowledge that sparks action.

The first observation is that the different certification schemes have very different views on what sustainability is and how it should be reached. That minimum social and environmental standards need to be reached is no discussion. But a prominent debate appears that some schemes want to provide a minimum price and market protection for small farmers, where other schemes seem to see this as an unfruitful intervention in an as such healthy market. The latter schemes merely aim to avoid unfruitful 'races to the bottom' by creating a level playing field of minimum social and environmental standards.[15]

The second observation is that action followed mainly from the interests and power position of actors in the chain. It hence seems that framing the problem as a knowledge-action gap is a mistake; we have to replace it by an analysis of actor positions, powers, and understand from there why certain actions are undertaken and not.

Links to Action

When we analyse the links to action more from a power and interest perspective, the following elements stand out:

[15] This discussion indeed played up within the team responsible for this book. My position with regard to the role of markets and minimum prices is the following. It is clear that free markets can play a negative role for sustainability (Scherhorn 2005), and they are too often are seen as a goal in themselves rather than a means to reach quality of life and foster human values (e.g. Speth 2008). And as shown in this case (but also other cases in this book), producers of agricultural commodities in developing countries usually do not have a powerful position in the value chain. This makes them vulnerable for being squeezed out. So developing certification systems for example that improve their market position is certainly needed. Yet, in my view such measures should not prevent normal market dynamics that for example correct over-capacities; in this context price guarantees should be carefully applied.

1. Certification schemes seem a good way to ensure basic standards for social and environmental responsible production, preventing the sheer race to the bottom that could result when market forces reign supreme in situations of overproduction. Even in cases where certification just demands compliance with local regulation it has an advantage. Certification implies an annual inspection of compliance with the certification rules, where state enforcement of legal rules in many of the producer countries is at best lax.
2. The consumer effectuates a surprisingly limited direct decision making power in the coffee value chain. Only in niche markets there is a real willingness to pay significant premiums for certified coffee, such as for fair trade and organic certified coffee. The added value of certification of a brand is hence low, and it makes little sense for mainstream producers and retailers to highlight the certification as part of their brand.[16]
3. Indirectly, however, consumer influence does play a role. Consumers in their role as citizens in the developed world tend to expect that reputable brands produce their products in a responsible way – a product attribute taken for granted just as that consumers expect that car does not rust, a refrigerator lasts at least10 years, etc.
4. In this sense, an interest with retailers, roasters and other brand owners exists to ensure a minimum quality and transparency in their supply chain. The Porter model suggests that these (next to the traders) are also the most dominant players in the coffee chain and in any case much more powerful than the many coffee producers.
5. Any certification scheme that has the ambition to become mainstream hence must win support of these dominant players. The alternative is, according to system innovation theory, to embark on a long-term strategy that is able to create an upheaval of the existing system and replaces it by a new configuration. This is hence no option for the short term, and given the power division in the system an unlikely future unless forces from outside the regime weaken the position of the current main players considerably.
6. The need to create support of the main players poses certain limitations on the type of demands that can be included in the certification schemes. The most visible one is that the two schemes that have the ambition to become mainstream – Utz Certified and Rainforest Alliance – pose no demands with regard to price setting. This, in many ways, reflects a fundamental discussion about how the economic system should work. One argument is that markets have in principle the basic function to regulate supply and demand. That new entrants (like Vietnam in the case of coffee) cause a major restructuring (at the expense of more expensive smallholders in Latin America) is not nice for those involved, but all in the game.[17] But the counter-argument is of course that where developed

[16] Worse, by highlighting the certificate producers may start to overshadow the elements of their brand that over time has of become distinctive value for the consumer vis a vis competition.

[17] For example, the massive 'cold sanitation' of the communities of farmers, bakers, and other middle class smallholders in Europe between 1945 and 1975 by the industrialization of production and the related need for scaling up, and the introduction of supermarkets.

countries created alternative means of existence or have developed (welfare-) systems to ease the pain for those affected, this is not the case in developing countries. In the coffee sector, small farmers end up drawing the short straw as being the least powerful in the value chain.

These findings can be understood well with the Porter and System Innovation models. Porter's model provides sound guidance to identify the most powerful actors in the production–consumption system, and who indeed has been the driving force for change. The system innovation model predicted that the more radical certification schemes would likely to stay niches. Does this mean, however, that such radical certification schemes have only limited value?. This is probably not true. As indicated, consumers in their role as citizens nowadays tend to expect that the goods they consume meet basic environmental and social standards. This has not always been the case. Though we did not explicitly research this in this paper, it is rather likely that idealistic or niche initiatives like Max Havelaar and 'third world shops', set up in the 1970s and 1980s, articulated societal awareness of the problem of unfair trade. They also showed the way how such unfair trade could be mitigated and transformed to fair trade. By articulating awareness, they most probably slowly changed the landscape: gradually but steadily it became unacceptable for the broad public that international commodity chains would be based on social and environmental exploitation. Mainstream players hence had to find an answer to deal with this expectation, and hence started to mainstream an adapted form of the niche solution, that fitted better to their situation. It is hence more than likely that the idealistic, niche schemes from the pioneering stages helped enormously to in two ways:

- A vehicle to change norms and values at landscape level
- An example for how mainstream practices could be changed

Leverage

As for creating leverage for further action the following can be said.

1. As for the future, it seems that fair trade and organic schemes are likely to stay speciality niches, unless a massive change in consumer behaviour occurs or other incentives are put in place.[18] The price premium of their products forces them to brand the label, and this seems only interesting for niche groups of consumers.
2. A point to be solved is that value chain logic seems to dictate that the 'value' of labels is much higher at the end of the value chain than at the farmer–trader interface. That farmers in some cases receive only a small part of the price premium may discredit labelling systems (see Box 10.1).

---

[18] For instance, in Switzerland organic farming became popular since under WTO rules this was the only way for the state to support farming (Belz 2004).

Box 10.1 Forces in the Value Chain

This piece below is excerpted from 'Go figure' by Tim Hartfort, Financial Times Magazine, 22 October 2005. The piece shows two things. First, it is an illuminating example how forces in the value chain determine added value: apparently an organic label is worth ten times more in the coffee shop than at the farmer–trader interface. Second, it also shows – in this case of coffee – how little in fact it costs for an improved fate of farmers with low power at the end of the value chain.

Costa, like most other coffee bars these days, offers "Fair Trade" coffee – theirs comes from a leading fair trade brand called Cafedirect. Cafedirect promises to offer good prices to coffee farmers in poor countries. Fair trade coffee associations make a promise to the producer, not the consumer. If you buy fair trade coffee, you are guaranteed that the producer will receive a good price. But there is no guarantee that you will receive a good price. For several years, customers who wished to support third-world farmers – and such customers are apparently not uncommon in London – were charged an extra 10p. They may have believed that the 10p went to the struggling coffee farmer. Almost none of it did. (...)

Cafedirect paid farmers a premium of between 40p and 55p per pound of coffee, and that premium was reflected in the price they charged to Costa. That relatively small premium can nearly double the income of a farmer in Guatemala, where the average income is less that $2,000 a year. But since the typical cappuccino is made with just a quarter-ounce of coffee beans, the premium paid to the farmer should translate into a cost increase of less than a penny a cup. (...)

Of the extra money that Costa charged, more than 90% did not reach the farmer. Cafedirect did not benefit, so unless using the fair trade coffee somehow increased Costa's costs hugely, the money was being added to profits. The truth is that fair trade coffee wholesalers could pay two, three or sometimes four times the market price for coffee in the developing world without adding anything noticeable to the production cost of a cappuccino. Because coffee beans make up such a small proportion of that cost customers might have concluded that the extra 10p was to cover the cost of the fair trade coffee, but they would have been wrong. A certain Undercover Economist made some inquiries and found that Costa worked out that the whole business gave the wrong impression, and at the end of 2004 began to offer fair trade coffee on request, without a price premium. (...)

But why had it been profitable to charge a higher mark-up on fair trade coffee than on normal coffee? Because fair trade coffee allowed Costa to find customers who were willing to pay a bit more if given a reason to do so. By ordering a fair trade cappuccino, you sent two messages to Costa. The first was: "I think that fair trade coffee is a product that should be supported." The second was: "I don't really mind paying a bit extra". Socially concerned citizens tend to be less careful with their cash in coffee bars, while unconcerned citizens tend to keep their eyes on the price. Perhaps another price list saying, "Cappuccino for the concerned £1.85. Cappuccino for the unconcerned £1.75?"

3. Utz Certified and Rainforest Alliance, despite their somewhat different backgrounds as initiated by business and NGOs, seem to follow a very similar strategy: playing out the interest of the roasters, retailers and other brand owner to ensure responsible behaviour in their supply chain, and offering a product tailored to this need. The vision that one or both of these labels will become mainstream, and coffee certification regular practice, seems not unrealistic.[19]
4. In theory, once 'basic labels' like Utz Certified and Rainforest Alliance have conquered the main market, these instruments could be used to strengthen standards with regard to the social and environmental aspects of sustainability. But it remains to be seen to what extent this will be possible: with these labels firmly rooted in the mainstream regime, they inevitably have to comply with the needs and expectations of the dominant players in the mainstream regime.
5. All labels discussed are voluntary systems, where of course an alternative approach could be to implement legal product or production standards with regard to coffee – *by an importing country or group of countries*. The institutional problem is that the WTO does not accept production standards in international trade.

The main leverage point for embarking on more far-reaching action as Utz Certified and Rainforest Alliance are doing on this moment hinges on the following elements:

1. There must be a convincing argument developed, that fair trade premiums do not form a distortion of markets, and that the fairness of a level playing field for minimum social and environmental standards is insufficient.
2. Once this vision is clear, it should be articulated by NGOs consumer organisations and pressure groups, which in turn force the main players in the chain to live up to such improved articulated norms and values.

Conclusion: A Judo Exercise by the Main Players

In sum, it seems that certification initiatives that have the ambition to become mainstream play a kind of 'judo' with the interests and make use of the forces that exist in the value chain. By carefully looking at the interests of the main players in the value chain, in part shaped by implicit consumer expectations about their behaviour, it seems possible to put a mainstream certification system in place that

[19] An interesting observation is that certification bodies and initiatives can be seen as competing in a market as well, developing a 'product' that needs to be 'accepted' by players in the value chain. A point not discussed in detail is that some roasters and other brand owners like Starbucks have taken some initiatives to arrange their own responsibility initiatives. However, it seems likely that they leave further development of certification systems to parties that see it as their core competence, as reflected by recent agreements between e.g. Utz Certified and Starbucks.

ensures minimum environmental and social standards. The environmental improvements can be significant, particularly if these schemes become a vehicle for integrated pest control practices. The more idealistic and radical certification schemes like Max Havelaar have a much tougher job to size a large market share, since they adhere to principles and practices that are difficult to accept for mainstream players. Examples are the strategy to brand the certificate (difficult for mainstream players since they brand their brand), and the practice to set minimum price levels (which in the view of mainstream players distorts the market, weakens competition and may lead to oversupply of coffee). At the same time, it is most likely that the mainstream certification schemes needed the example of the idealistic and radical certification schemes:

- Schemes like Max Havelaar raised awareness about the issue of "fair trade", and paved the way for a change of (implicit) norms and values at landscape level about acceptable business behaviour
- Schemes like Max Havelaar tried and tested at a solution how socially and environmentally responsible businesses could be implemented

In sum, the mainstreamed certification schemes may play the game slightly different as the niche schemes, and may seem now more 'successful'. It is however difficult to understand their current achievements without taking the pioneering role of the niche schemes into account.

Finally, the example has to be described as a success in the field of sustainable production rather than sustainable consumption. The role of the consumer as a driving factor is indirect. The consumer in its role of citizen is mainly a carrier of (implicit and tacit) norms and values about sustainable and responsible business, and the actors in the value chain seem to respond to this demand. For the consumer/citizen as such, there is not any behavioural change involved. The example also has to be characterised as an incremental rather than a radical change to SCP. Although the case shows interesting examples of (normative) change at landscape level, and new initiatives in niches, the structure of the production–consumption system (or socio-technical regime) had not changed fundamentally. Indeed, once could say that idealistic and radical niche approaches (fair trade schemes like Max Havelaar) have been adopted and modified by regime players in a way that does not ask for radical change, as an answer to new demands about responsible business that emerged at landscape level.

References

Berkhout F, Smith A, Stirling A (2004) Socio-technological regimes and transition contexts. In: Elzen et al (2004) op cit

Belz FM (2004) A transition towards sustainability in the Swiss agri-food chain (1970–2000): using and improving the multi-level perspective. In: Elzen et al (2004) op cit

Consumers International (2005) From bean to cup: how consumer choice impacts upon coffee producers and the environment. Consumers International, London

Correljé A, Verbong G (2004) The transition from coal to gas: radical change of the Dutch gas system. In: Elzen et al (2004) op cit

Davies A, Brady T, Tang P with Hobday M, Rush H, Gann D (2003) Delivering integrated solutions. SPRU, Brighton, p 34
Dosi G (1982) Technological paradigms and technological trajectories. Res Policy 11:147–162
Elzen B, Geels FW, Green K (2004) System innovation and the transition to sustainability. Edward Elgar Publishers, Cheltenham
Fischer-Kowalski, M., H. Haberl and H. Payer (1994). A plethora of paradigms: outlining an informatin system on physical exchanges between the economy and nature. In: R. Ayres and U.E. Simonis, Industrial Metabolism. Restructuring for Sustainable Development. Tokyo, Japan: United Nations University Press
Geels FW (2002) From sectoral systems of innovations to socio-technical systems. Insights about dynamics and change from sociology and institutional theory. Res Policy 33:897–920
Hartford T (2005) The undercover economist. Oxford University Press, New York
Hertwich, E. 2005. Life cycle approaches to sustainable consumption: A critical review. Env. Sc. Tech., Vol. 39, No 13, p4673
Kemp R, Rotmans J (2004) Managing the transition to sustainable mobility. In: Elzen et al (2004) op cit
Munasinghe and W. Shearer (eds.) (1995), Defining and Measuring Sustainability, The Biogenophysical Foundations. Washington DC, US: United Nations University and the World Bank
Nelson, R.R. and S. Winter (1982) An Evolutionary Theory of Economic Change, Cambridge, MA: Harvard University Press
Rip A, Kemp R (1998) Technology change. In: Rayner S, Malone EL (eds) Human choice and climate change, vol 2. Batelle Press, Columbus, OH, pp 327–399
Prahalad, C.K. (2004). The fortune at the Bottom of the Pyramid. Eradicating Poverty Through Profits. Wharton School Publishing
Porter M (1985) Competitive advantage. Creating and sustaining superior performance. The Free Press, New York
Potts (2003) Building a sustainable coffee sector using market based approaches: the role of multi-stakeholder co-operation background paper for the meeting of the sustainable commodity initiative, UNCTAD-IISD, New York. www.ico.org/documents/sustain1.pdf. Cited 15 July 2006
Ponte S (2002) Standards, trade and equity: lessons from the specialty coffee industry. Centre for Development Research, Working Paper 02.13, Copenhagen, Denmark
Riele H te, Duifhuizen SAM, Hotte M, Zijlstra G, Sengers MAG (2000) Transities: Kunnen drie mensen de wereld doen omslaan? (Can three people change the world?) Environment Ministry, Publication Series Environmental Strategy, den Haag, Netherlands
Scherhorn, G. (2005). Sustainability, consumer sovereignty and the concept of the market. In: K.G. Grunert and J. Thøgersen: Consumers, Policy and the Environment. A Tribute to Folke Ölander. New York, US: Springer
Slob B, Oldenziel J (2003) Coffee and codes. Overview of codes of conduct and the ethical trade initiatives in the coffee sector. SOMO, Amsterdam
Speth, JG (2008) The bridge at the edge of the world. Capitalism, the environment, and crossing from crisis to sustainability. Yale University Press, New Haven, CT
Tukker A, Huppes G, Suh S, Geerken T, Jansen B, Nielsen P (2006) Environmental impacts of products. Literature review and input–output analyses on 500 product groupings for the EU's Integrated Product Policy. ESTO/IPTS, Sevilla
Tukker A, Tischner U (2006) New business for Old Europe. Product service development, sustainability and competitiveness. Greenleaf Publishing, Sheffield
Tukker A (2008) Sustainability: a multi-intepretable nation. In: Tukker A, Charter M, Vezzoli C, Sto E, Munch Andersen M (eds) System innovation for sustainability 1: perspectives on radical changes to sustainable consumption and production, Chapter 2. Greenleaf Publishing, Sheffield, UK
UN (2002) Plan of implementation of world summit on sustainable development. UN, New York
Utz Certified (2006a) Utz Kapeh code of conduct. Utz Certified, Amsterdam, Netherlands
Utz Certified (2006b) Utz Kapeh Newsletter March 2006. Utz Certified, Amsterdam, Netherlands
Wise R, Baumgartner P (1999) Go downstream – the new profit imperative in manufacturing. Harv Bus Rev 77. Harvard University Press, Cambridge, MA

Chapter 11
Production and Consumption of Tourist Landscapes in Coastal Areas: Case Study of Tourism in Malaysia

Ooi Giok Ling

Introduction

The development of both natural and built heritage for tourism has been part of national agendas particularly in Southeast Asia. Coastal resort developments have sprung up throughout the region. Not only do these tourist attractions seek to compete with those that have been developed among other countries in the region but many have been created to 'outdo' other similar resort developments within the same country. This appears to be especially characteristic of coastal resort developments in Malaysia. The following proposed case-study considers the development of coastal resorts in Malaysia with a focus on one – Pulau Langkawi – along the north-western coast of Malaysia. Pulau Langkawi is one among the many island resorts that have been developed for tourism. Along the west coast of Peninsular Malaysia, the most important resorts include Pulau Langkawi, Penang Island and Pangkor Island. In the east coast area, island resorts include Pulau Tioman, the Perhentian Islands and Pulau Redang.

Island resort development dates from 2,000 years ago when the Romans first developed the Isle of Capri as a holiday destination (Conlin and Baum 1995). 'Confronted with a limited number of development options because of their isolation, small physical size and lack of resources, many islands regard tourism as a panacea for their economic difficulties despite the pervasive negative economic, social and environmental impacts that are often associated with island tourism' (Tan 2000, p. 43). In the discussion, the dilemmas of development that have emerged concern the failure to reconcile local heritage, both natural and built, with the construction of tourist attractions. While there is a strong effort to 'sell' the natural beauty of the island resort, the developments are damaging beaches and other nature areas. Little consciousness has also been displayed by planning officials of the links that exist between the biodiversity and wildlife to be found on the islands and natural heritage generally. The prospects for a more sustainable agenda in tourism development therefore, appear rather bleak.

O.G. Ling
National Institute of Education, Nanyang Technological University, Singapore
e-mail: giokling.ooi@nie.edu.sg

L. Lebel et al. (eds.), *Sustainable Production Consumption Systems: Knowledge, Engagement and Practice,* DOI 10.1007/978-90-481-3090-0_11,

Relevant authorities in Malaysia seem to be focused on attracting as many consumers as possible to the products, that is, the islands with their natural attractions since the resources sought by tourists – sun, sea and sand – are regarded as renewable, abundant and also inexpensive or cost relatively little to provide (Milne 1990).

In research, there has been relatively less attention paid to tourist attractions compared to the transport, accommodation and tour operator components of the tourism sector (Wu and Wall 2005; Swarbrooke 2002). While there is agreement that tourism competitiveness can be increased by improving the environmental quality of the tourist destination through sound environmental management practices (Hassan 2000; Huybers and Bennett 2003; Mihalic 1999), there are few empirical studies and confirmatory examples at both macro and micro levels (Wu and Wall 2005). Indeed, a framework to demonstrate the way in which sound environmental practices can enhance the competitiveness of a nature-based tourism destination has been proposed (Huybers and Bennett 2003). Environmental management at a nature-based destination would have to include a public and a private component. In the model illustrated in Fig. 11.1, the costs of sound environmental management with a mix of government or public sector regulations as well as voluntary regulations are acknowledged to raise costs which might affect competitiveness of the tourist destinations negatively. Countering such a negative outcome however, would be the growth in demand and the shift to a more virtuous cycle of high environmental quality, the sustainability of the tourism development and continued high demand as a result.

Developing the Production Consumption System of Island Resorts for Tourism

Malaysia has been relatively late in focusing on tourism development compared to other countries in Southeast Asia including neighboring Singapore. The quinquennial Malaysia Plans have emphasized instead the primary production and industrial

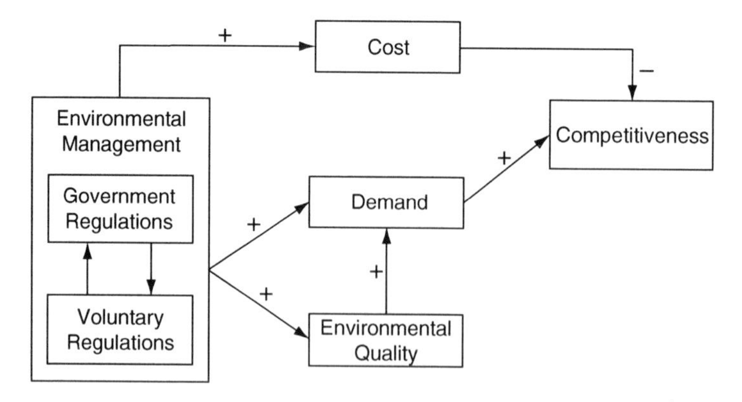

Fig. 11.1 Environmental management and competitiveness of a nature-based tourism attraction Huybers and Bennett 2003)

sectors right up to the 1980s. The Ministry of Culture and Tourism was established in 1987 and subsequently became the Ministry of Culture, Arts and Tourism in 1990. Investments in the tourism sector have however been growing rapidly since. In the Second Malaysia Plan of 1971–1975, investments in tourism totaled some $17.2 million Malaysian ringgit (US$4.7 million) but by the Seventh Malaysia Plan of 1991–1995, the amount was $966 million Malaysian ringgit (US$267 million) (Hall and Page 2000, p. 145).

In 1980, total arrivals in Malaysia as a whole were about 2.2 million, but these rose to 3.1 million in 1985 and 7.4 million in 1990 – an increase of 230% over the 10 years (Lee 1996). By 2000, annual tourist arrivals had reached 10.2 million and tourism receipts had reached some $13.4 million Malaysian ringgit (US$3.7 million) in 1999 (Malaysia Tourism Promotion Board 2001). While the majority of the tourists originally came from Southeast Asia, that is, the more affluent countries of Singapore and Thailand, tourists from other countries like Japan, Australia, the United States, India, the United Kingdom and Hong Kong as well as subsequently Taiwan, South Korea and West German, have grown in numbers since 1980. By 1989, tourists from the Middle East and a number of other European countries had also increased.

The tourism industry in Malaysia has encouraged the development of numerous integrated island resorts including golf courses which have generally been perceived as being largely incompatible with the ecology of small islands (Shamsul and Teh 2000, p. 213). Malaysia's first Prime Minister, Tunku Abdul Rahman, urged the development of one of these islands, Pulau Langkawi, into a tourist destination upon which the Public Works Department cleared a rubber plantation for a nine-hole golf course that was declared open by the Sultan of Kedah in 1970. The course was eventually abandoned both because of a lack of funds to maintain the course as well as public interest in using it (Shamsul and Teh 2000, p. 219). When the government revived its objective of developing Pulau Langkawi into a major tourist destination in 1989, the course was then upgraded and converted into an 18-hole facility.

Pulau Langkawi is the largest among a breathtakingly beautiful group of 99 islands off the coast of northwestern Malaysia. Since the 1990s, there has been an ambitious tourism development programme launched in Pulau Langkawi including the establishment of the Langkawi Development Authority that reported directly to the prime minister's office. Langkawi's key selling points have been the pristine conditions of its beaches and the sea together with the sun and good tropical weather. The karst landscape in the form of rocky outcrops and low hills lend themselves to the idyllic natural scenery that has formed the main theme for the marketing of the island to tourists – its mist-shrouded landscapes and their collection of myths and legends.

The country to which Pulau Langkawi belongs, Malaysia, had worked to earn its place as an export platform by offering the semiconductor industry among other incentives, long tax holidays (Greider 1997). Dr Mahathir Mohammed was the prime minister of Malaysia between 1981 and 2003. The Mahathir government was bent on securing its vision of a developed country status for Malaysia by 2020 – *Wawasan* 2020. The vision included the quadrupling of per capita income by 2020 and rapid growth of 7% annually. If the plan sounded rather ambitious, the development projects that the state sector embarked on was certainly evidence that Wawasan 2020 was not going to be just empty talk.

In the context of tourism development and cultural consumption, local identities have been incorporated into economic projects of the nation-state. The post-modernist argument of hyper-commodification involves the commodification of all areas of social life (Crook et al. 1992). Such a commodification process of culture and in turn its consumption reflects the concerns about the erosion of the distinction between authentic and inauthentic culture (Baudrillard 1983).

The penetration of capitalism into many areas of leisure encourages consumption and this according to Harvey (1990) has resulted in the ferment, instability and fleeting qualities of a postmodern aesthetic that celebrates ephemerality, spectacle, fashion and commodification of culture (Haralambos and Holborn 2004). Like "McDonaldization" which leads to the substitution of illusion for reality according to Ritzer (1996), so the packaged tour and international villages in amusement or theme parks like Disneyland create 'pseudo events' that are staged but pretend to give people the real experience.

In this process of manufacturing attractions for tourists, there has been the 'serial production of world trade centres, waterfront developments, post modern shopping malls' (Harvey 1989, p. 10). Most evident in, as well as among cities, '… this competition location game …' promotes the marketing of city and region and '… their imageability becomes the new selling point'. Consequently, spatial design codes and architectural pattern languages become increasingly important in selling the look of an upmarket, upbeat environment. In this marketing war, 'style of life, visualized and represented in spaces of conspicuous consumption, become important assets …' (Boyer 1992, p. 193).

This case-study examines the agendas both economic and political as well as the deleterious effects of the Malaysian production and consumption system for the global tourism market. In discussing the construction of landscapes for tourist consumption, the case study intends to highlight the irrevocability of the commodification of 'place' and nature to be found. The destruction of not only nature upon which the tourism sector relies particularly in the case of Pulau Langkawi but also the cultural changes that have occurred provide evidence of the lack of sustainability of the development agenda for the island and tourism as a whole in Malaysia.

Prospects of an Ecological Agenda for Malaysian Coastal Tourism

In Malaysia, official attempts in tourism development as an integral part of the restructuring of the national economy date back to 1985. With the declaration of Visit Malaysia Year in 1990, the federal and various state governments became preoccupied with promoting and developing the tourist industry in the country (Tan 1992, p. 264). Research however has highlighted both environmental impacts like beach pollution from untreated sewage being discharged by hotels into the sea and changes in local cultures because of tourism trade and services introduced to the localities since tourism accelerates cultural changes (Mathieson and Wall 1982; Iyer 2000).

With its aim of achieving developed country status by 2020, Malaysia's strategy features both industrial production as well as services such as trade and tourism (Government of Malaysia 1991). Not surprisingly, tourism appears to be the third largest economic sector in Malaysia and in 1995, earnings of this sector totaled some RM10.5 billion (estimated US$3 billion). This revenue from the tourism sector was second only to both manufacturing and the oil palm industries (Malaysian Tourism Promotion Board 1998).

Observers of the development of coastal resorts in Malaysia such as Pulau Tioman point out that like other parts of the country, the tourism industry in Tioman Island is largely 'hardware-driven'. Emphasis is put on physical development to provide accommodation ranging from the large international resort to village-level chalets, public amenities in transport and communication, water and electricity supplies, recreational space for golfing and swimming, and service-oriented facilities such as restaurants and handicraft shops. Development may take the form of large-scale, capital-intensive or small-scale projects. Among the former are the international beach resorts and jetties' (Voon 2000). Development has taken precedence over rural interests with the displacement of entire settlements by beach resorts as well as the collapse of activities like fishing and farming. Agricultural activities are gradually abandoned like fishing because providing transport services for tourism has proven more lucrative (Voon 2000). This is happening in Pulau Langkawi where farming communities continue to grow rice the way they have done for the last century or so.

Promotion of Pulau Langkawi to compete with nearby rivals like the islands of Penang in Malaysia and Phuket in neighboring Thailand has been keen and the island was accorded free port status in 1987. The government has invested some RM500 million (US$130 million) in infrastructure while more than 100 projects by the private sector developers worth some RM1 billion (US$260 million) have been approved (Teh and Ong 2000, p. 277). Some 7,500 hotel rooms have been planned to be built by 2005 (Langkawi Development Authority 1990). By 1995, the islands of Pulau Langkawi were receiving some 1.7 million tourists (Malaysian Tourism Promotion Board 1998). Beaches were the main selling point; one of the most intensively developed bays, Pantai Chenang, on the eastern side of the largest island, is already facing erosion with resort and chalet owners often having to find solutions to protect their properties (Teh and Ong 2000).

Langkawi lies off the coast of the Malaysian state of Kedah in the northwestern part of the country. The state claims it has 110 islands of which 99 are in Langkawi alone. Islands in Malaysia share a rich history that is often neglected in the headlong rush to exploit them for tourism development. Many of the islands have been homes to seafarers, pirates and fishermen as well as even detention centers and places for quarantine. Refugees have been accommodated on islands. Most islands have been popular destinations for anglers and hobby fishers (Teh 2000).

Land reclamation works have been part of the resort development in Pulau Langkawi to provide land for an airfield to allow international flights to the island as well as other commercial developments. In the meantime, other construction works have been launched and already account for erosion seen at the beaches that have been the most popular among tourists. Ironically, the tourism in Langkawi still relies heavily on its beaches. Many of the tourist attractions that have been developed have 'failed' while others are in the process of being 're-invented' to boost tourist numbers. There appears to be little in the state agenda to flag the review of a tourism development pathway which is unsustainable and urgently in need of effort to bring ecological and other costs into the equation.

Sustainability Issues in the Production and Consumption Systems of Island Tourism Landscapes

More than perhaps in other Southeast Asian countries, tourism development in Malaysia has been a strategic tool wielded by the state sector for bringing modernization and modernity to territories that are considered 'backwaters' and in need of 'development'. Malaysia's Wawasan 2020 has sought to develop projects that are essentially about national identity and nationalistic agendas. These projects include many tourism development initiatives particularly in the country's many offshore islands.

The development of tourism in islands such as Pulau Langkawi certainly thrusts both the island and its population into the global stage and the competition for both tourists and tourist dollars. Such a developmental trajectory follows a familiar formula – the existing agricultural or rural landscapes are razed to the ground to make way for themed landscapes that are expected to attract tourists. These are landscapes adapted from theme parks elsewhere as well as 'cultural' inventions based more or less loosely on the cultural heritage and identity of the islands in which these landscapes are being developed. Indeed, the development of different theme parks and tourist attractions has effectively led to sprawl with many of these developed sites located outside of the town centre and away from the international airport.

To facilitate travel on the island of Pulau Langkawi, there is car hire or rental services as well as taxis operated by individual owners of mini-vans. The island's residents mainly use private modes of transport – motorcycles and cars. All these vehicles plus the fuel to run them are imported to Pulau Langkawi. The volume of traffic has been such that it is often not practical to walk even within the town centre. Not only is there no provision for pedestrian walkways, land-use is not planned to allow conveniences to be clustered in sites that are accessible to pedestrians. The public bus transport services were terminated because of lack of usage and costs involved in running the services.

The development of the infrastructure for tourism does not appear to have considered a more sustainable form of transport for the main island of Pulau

Langkawi. Land use development for the variety of tourist attractions appears to have by-passed the different agricultural activities that have been an inherent part of the island's economy – rubber plantations, padi or rice farming and fishing. These are traditional economic activities that a nation-state with its ambition of becoming a developed country in 2020 seemingly would not want to be associated with.

Characterizing the unsustainable production of tourist landscapes are two tourism projects that are now largely defunct or going through yet another make-over as tourist attractions. First is the creation of the now largely defunct theme park including a museum to exhibit to tourists the growing processes of rice and its history. This theme park includes plots of rice fields among other facilities including eateries and grounds for parties. Today, the theme park sees less than a handful of tourists although entry is now free. Located close to beach resorts and hotels, the theme park was meant to provide a convenient leisure activity to tourists so that they would not have to take the longer trip to the 'real' rice fields dotting the entire island together with village settlements and farmers. Brown eagles – the *lang* – which has given the island its name of Pulau Langkawi – circle the rice fields possibly attracted by the small birds and other prey to be found in them. None of these are on the tourists' itineraries however. There is instead a giant cement sculpture of a brown eagle in the square which has been built off the ferry terminal as part of the tourism promotion.

Second, the shrine of the legendary princess, Mahsuri, who was wrongfully put to death, has been a tourist attraction that has undergone several rounds of re-making. The shrine itself has been the site of pilgrimages for both Buddhists as well as other Malaysians generally who think of the well-loved legend of the pure and beautiful princess as their own. For the majority of the domestic visitors, the shrine where the princess is reportedly buried is the key attraction in Langkawi. This may explain the importance of the shrine as a tourist attraction to the local and national tourism promotion authorities. A theme park has not been built at the site of the shrine but a tombstone has been placed in a museum and Buddhist worshippers who used to place incense at the shrine have since been discouraged from following this practice. The theme park owners explained that this practice is not appropriate in a Muslim society.

While the legend furnishes the rationale for the theme park, the shrine is now besieged with other developments including a stage for traditional Malay cultural performances. There are traditional Malay homes that have been built to add to the attractions with space for the exhibition of cake-making and other cultural practices tailor-made for tourism. Shopping facilities and sale of souvenirs have been part of the new developments put in place in the theme park.

The side-stepping of local cultures as well as the needs of the residential population in island resorts has been a familiar and iterative outcome of island resort development. At Penang Island, local devotees no longer go to the temples that are on the tourist itineraries because of the perception that the tourists have chased away the 'spiritual presence' of the gods. Conservation by the state has made the temples 'too colourful, like a zoo for the gods' (Teo 2003, p. 558). Indeed, locals are actually discriminated against and denied access at beach resorts because of the cordoning off of beaches for the exclusive use of hotels developed throughout the islands (Teh 2000; Teo 2003).

Like the proliferation of golf courses developed in the island resorts, the developers of tourist attractions on the islands have not worked collectively for the benefit of the islands – as deforestation, air and river as well as water pollution and high bacterial content in water at beaches among issues flagged by civil society groups highlight (Teo 2003, p. 559). While the marketing of the beach resort developments promise sun-kissed beaches and golden sunsets, private sector developers like the state development authorities overseeing tourism development, have generally neglected the pollution along the beaches as well as the soil erosion at sites abutting the beaches.

Conclusion

Indeed, the 'products' into which islands of Malaysia have been transformed for tourism are major sites of contestation between state and society, local and tourists' needs as well as development and conservation. Not only are communities being displaced to make way for tourism development projects but local islanders lost access to their beaches until protests and demonstrations were mounted after which some limited access was granted. Furthermore, much of the focus in the development of island infrastructure has been on facilities to give tourists access to the islands as well as services such as, accommodation, eateries, local transport and shopping. Basic needs of islanders can be wholly neglected. Ultimately, demand for water and other resources rise in part because of the neglect of environmental protection measures to prevent water as well as other forms of pollution. Indeed, future and further development of island resorts in Malaysia, requires a collective review and re-thinking of the development goals in this production consumption system including and benefits that can be shared with the local population.

References

Baudrillard J (1983) Simulations. Semiotext(e), New York
Boyer MC (1992) Cities for sale: merchandising history at South Street Seaport. In: Sorkin M (ed) Variations on a theme park: the new American city and the end of public space. Hill & Wang, New York, pp 181–204
Conlin M, Baum T (eds) (1995) Island tourism: an introduction in island tourism: management principles and practice. Wiley, Chichester, pp 3–14
Crook S, Pakulski J, Waters M (1992) Postmodernisation: changes in advanced society. Sage, London
Government of Malaysia (1991) The Second Outline Perspective Plan 1991–2000. National Printing Department, Kuala Lumpur
Greider W (1997) One world ready or not. Simon & Schuster, New York
Haralambos M, Holborn M (2004) Sociology – themes and perspectives. Collins, London
Harvey D (1989) The condition of postmodernity – an inquiry into the origins of cultural change. Blackwell, Oxford
Hassan S (2000) Determinants of market competitiveness in an environmentally sustainable tourism industry. J Travel Res 38:239–45

Huybers T, Bennett J (2003) Environmental management and the competitiveness of nature-based tourism destinations. Environ Resour Econ 24:213–33
Hall CM, Page S (eds) (2000) Tourism in South and Southeast Asia: issues and cases. Butterworth, Heinemann, Oxford
Iyer P (2000) Bali: on prospero's isle/the Philippines: born in the USA. In: Lechner FJ, Boli J (eds) The globalization reader. Blackwell, Oxford, pp 111–117
Langkawi Development Authority (1990) Langkawi Structure Plan 1985–2005
Lee BT (1996) Emerging urban trends and the globalizing economy in Malaysia. In: Lo F-C, Yeung Y-M (eds) Emerging world cities in Pacific Asia. United Nations University Press, Tokyo
Malaysian Tourism Promotion Board (1998) Malaysian Tourism, July–August
Mathieson A, Wall G (1982) Tourism: economic, physical and social impacts. Longman, London
Mihalic T (1999) Environmental management of a tourist destination: a factor of tourism competitiveness. Tourism Manag 21:65–78
Milne S (1990) The economic impact of tourism in Tonga. Pacific Viewpoint 31(1):24–43
Ritzer G (1996) The McDonaldization of society. Pine Forge Press, London
Shamsul B, Teh TS (2000) Island golf course development in Malaysia: an environmental appraisal. In: Teh TS (ed) Islands of Malaysia – issues and challenges, IRPA R & D Project. University of Malaya, Kuala Lumpur, pp 213–234
Swarbrooke J (2002) The development and management of visitor attractions. Butterworth-Heinemann, London
Tan, PK (1992) Tourism in Penang: its impacts and implications. In: Voon PK, Tunku SB (eds) The view from within – geographical essays on Malaysia and Southeast Asia, Kuala Lumpur, Malaysia. Malayan J Trop Geogr 263–278
Tan WH (2001) Island tourism development: a case study of the Perhentian Islands, Terengganu. In: Teh TS (ed) Islands of Malaysia – issues and challenges, IRPA R&D Project. University of Malaya, Kuala Lumpur, pp 43–58
Teo, P (2003) 'The Limits of Imagineering: a Case Study of Penang', International Journal of Urban and Regional Research, 27:3, 545–563
Teh TS, Ong A (2000) An analysis of construction setback lines of beach resorts: with special reference to Langkawi Island. In: Teh TS (ed) Islands of Malaysia – issues and challenges, IRPA R&D Project. University of Malaya, Kuala Lumpur, pp 273–295
Teh TS, Tengku SB, Ooi CH (2000) Future of Pulau Layang-Layang: Man-made Paradise or Ecological Disaster? In: Teh TS (ed) Islands of Malaysia – Issues and Challenges, IRPA R & D Project, University of Malaya, Kuala Lumpur, pp185-202
Voon PK (2000) Tourism and the environment: the case of Tioman Island, Malaysia. In: Teh TS (ed) Islands of Malaysia – issues and challenges, IRPA R&D Project. University of Malaya, Kuala Lumpur, pp 151–170
Wu W, Wall G (2005) Environmental management, environmental image and competitive tourist attraction. J Sustain Tourism 13(6):617–635

Chapter 12
Enhancing Sustainable Tourism in Thailand: A Policy Perspective

Hannarong Shamsub

Introduction

The goal of tourism development is to bring foreign currency into the national economy and to distribute income at the local level. In Thailand, the government sets an annual tourism target of the number of foreign visitors. Although the target seems to keep increasing each year along with rising advertising budgets, the number of visitors has not kept pace. For example, the target for 2006 was 14 million visitors but the actual number was 12 million. Conventionally, to attract international tourists, policy makers in Thailand prefer spending more on advertising than evaluating the tourism development process. The chapter identifies both sustainable and unsustainable practices in the tourism industry in Thailand and proposes policies to move the industry in new sustainable direction.

Tourism resources have been exploited to produce short-term profits rather than long-run gains for the entire economy and local development. This has resulted in the environmental or cultural degradation of many tourist attractions that have subsequently closed down or lost popularity. Consider the following example.

Around 30 years ago, Pattaya beach was a paradise for both international and domestic tourists. But uncontrolled tourism development resulted in Pattaya beach being transformed from a beautiful beach which attracted high-value guests to a heavily polluted and overly exploited "cowboy" town that attracts low-price package tours and hosts international criminals. The sea water is now heavily polluted by untreated waste water released from tourism businesses as well as residential properties. The water has become so dirty that swimmers have developed skin rashes (Tangwisutijit 2004). When Pattaya lost its popularity, Jom Thien, another newly developed beach located a few miles North of Pattaya emerged. Now Pattaya is ranked as a cleaner beach than Jom Thien, which had the worst sea water quality among the 14 most popular beaches in Thailand (Tangwisutijit 2004). Poor planning is a recurrent feature of unsustainable practices in the Thai tourism industry.

H. Shamsub
Lecturer-MBA Programs, RMIT International University, Ho Chi Minh City, Vietnam
e-mail: hannarong.shamsub@rmit.edu.vn

L. Lebel et al. (eds.), *Sustainable Production Consumption Systems: Knowledge, Engagement and Practice*, DOI 10.1007/978-90-481-3090-0_12,

Sustainable Tourism

The World Tourism Organization (WTO) defines sustainable tourism as "tourism which leads to management of all resources in such a way that economic, social and aesthetic needs can be filled while maintaining cultural integrity, essential ecological processes, biological diversity and life support systems" (McKercher 2003). In addition, I believe sustainable tourism development should follow four principles: inter-generation equity, intra-generation-equity, environmental protection, and public participation.

This chapter draws on 1 year of research on practices in Thailand's tourism industry. The research utilizes the sustainable production and consumption systems framework to uncover sustainable practices used by various tourism stakeholders: local residents residing in the tourism areas, tourists, tourism businesses, tourism site operators, and various government agencies. Several research methods were used including: structured and structured interviews, observation, visitor exit survey, and secondary data collected from printed media. Structured interviews were used to investigate local residents' perception toward the impacts of tourism. Unstructured interviews were used to uncover sustainable business practices in tourism businesses. Data on visitors' behavior were collected using observation and visitor exit surveys. Data from printed media were used to substantiate arguments or opinions received from interviews. Although this section emphasizes international tourists, domestic tourists are mentioned periodically for comparison purposes.

To cover various types of tourism, eight main tourism sites in Thailand were visited. Two sites were selected for each type of tourism. Ayutthaya Historical Park in Ayutthaya province and Phi Mai Castle in Nakorn Ratchasima province represent cultural tourism, while Doi Inthanon in Chiang Mai and Kao Yai National Park in Saraburi province represent ecotourism. Phuket and Phang Nga are representatives of sea-sand-sun (3s) tourism. Yao Noi Island in Southern Thailand and Huay Pooling villages on Doi Pui mountain, Northern Thailand, are representatives of community-based tourism (CBT).

The chapter is organized as follows. The next section discusses what local residents gain and lose from tourism. This is followed by sections on sustainable business practices, services provided by tourism sites, sustainable tourist behavior, and sustainable tourism as a public policy. The concluding section offers recommendations for policy.

What People Gain (Lose) from Tourism

Tourism uses many types of resources including natural and cultural. Use of these resources creates impacts, both positive and negative, on the lives and livelihoods of people, especially those who live within and near tourism sites. How people perceive the impacts of tourism on their lives is one indicator of the success of tourism.

Villagers I interviewed had mixed perceptions of how tourism affects their communities and lives. Most villagers perceive that tourism helps improve the local economy, livelihood and living conditions. The following reasons illustrate how tourism makes those things happen: (1) tourism increases income for existing local businesses, (2) creates business opportunities for impoverished people to work in the informal sectors in the form of roadside and mobile vending, (3) generates tourism-related careers, (4) creates supplemental income from part-time jobs in tourism, (5) makes some labour market drop-outs to become active and (6) increases revenue for local governments. While most people would agree that tourism generates positive economic impacts to local communities, many local residents also point out worsened income distribution as the main problem associated with tourism. Many perceive tourism as benefiting mainly big businesses.

The most disliked impacts of tourism split nearly evenly between environmental and cultural degradation. Water pollution was perceived as the most serious threat of all environmental threats to sustainable tourism and the causes of the problems include irresponsible business behavior, lack of adequate treatment facilities provided by governments, and lack of enforcement of environmental and building laws. These three problems are intertwined. For example, hotels having 80 rooms or more are required to have water treatment facilities. However, some businesses exploit this legal loophole to build a hotel of 158 rooms, using two separate licenses, each of which having 79 rooms. Some hotels install water treatment facilities, but turn them on only when environmental officers visit the sites. Some use their cozy relationship with the relevant authorities to bypass regular inspection.

Small establishments, for example, hotels having less than 80 rooms, are not required to have water treatment facilities. Waste water from these establishments is released to the city's sewage systems and ends up at the city's treatment facilities. However, many cities such as Patong and Krata Kraron in Phuket, do not have adequate number of treatment facilities to keep pace with increasing demand. This is also due to the fact that local governments are not fiscally independent, relying on capital from the central government.

The socio-cultural impacts perceived by villagers arise from differences in culture between themselves and visitors. Due to differences in culture, some tourist behavior is perceived by residents residing in the areas as culturally inappropriate. Such behavior includes climbing the stupas or Buddha statues, sunbathing naked, males walking topless in the street, and females wearing provocative and skimpy clothing. The main concerns of residents are that their youngsters may mimic the visitors' inappropriate behavior. On the positive side, many villagers feel that tourism motivates them to preserve local traditions, stimulates cultural exchange, and encourages cooperation among local residents. An interesting finding worth mentioning is that some minority groups think that tourism gives them opportunities to purge any negative perceptions of their communities. For example, many Thai people have the mistaken perception that the residents of the Thai Mai (formerly Morgan) communities located in some southern provinces are "dirty" because they live by the sea and have simple dwellings. Despite this negative perception, Morgan

Fig. 12.1 Dwellings of a Thai Mai (Morgan) community in Phang Nga province; last picture on the right shows some local products sold by the community

people always welcome outside visitors to show that their simple way of life near the sea is not tantamount to "dirty" (Fig. 12.1).

Business Practices

The tourism industry consists of many business sub-sectors. In general, business practices for each sub-sector are far from being sustainable. There are many obstacles for tourism businesses to become sustainable. Those include the market structure of the developed (tourist-generating) countries, global competition among global tourism-related businesses, attitude toward sustainable tourism, and the effect of global competition on the local tourism industry.

Global Tourism Competition

At the global level, tourism businesses are highly competitive. The tour operator and travel agent sectors in developed tourist, generating countries including the USA, UK, France and Italy, are characterized by high concentration ratios and high level of entry and exit.

The high level of market concentration is a result of integration among firms within the same and also cross sectors. It remains controversial whether such integration promotes greater efficiency. However, some scholars point out that the larger size resulting from integration is associated with greater concentration and market power in the origin country, higher volume sales and increased bargaining power with hoteliers in developing country destinations, who eventually tend to receive relatively low income for each room occupied by clients of the tour operators (Sinclair 1998).

Tour operators have different attitudes toward sustainable tourism (Curtin and Busty 1999). In the UK, specialist operators are more concerned with tourism sustainability than mass operators. Many tourism businesses in some European countries believe that they are powerless in making changes. Many firms are unwilling to adopt sustainable practices, citing the lack of demand for sustainable tourism. In addition, some businesses perceive sustainable tourism as academic and irrelevant. Some hotels do not implement sustainability practices because they don't want to appear to "preach" to high-spending guests. In order to implement sustainable practices, some businesses feel that they need "expert assistance" and financial support from public funds to compensate for extra costs involved when introducing sustainable practices.

High competition in the global tourism market sometimes can have adverse effects on the tourism supply chain. This is illustrated by the rise of "below-cost pricing" as can be seen in the case of the "zero-dollar tour".

Zero-Dollar Tour

The zero-dollar tour (*Tour Soon Rien* in Thai language) brings in visitors from China and India (and is presently also expanding to some European countries such as France) to Thailand for a 5 day package for only a couple of 100 dollar. It operates in reverse of a normal group-tour operating process. Inbound tour operators in the host countries such as Thailand sell package tours to outbound tour operators in an originating country such as China. The offer price of a normal package tour usually covers operating costs including mark-up. But the package price for a budget or zero-dollar tour offered by local inbound tour operators is below operating costs (starting at US$80 about 10 years ago).

When the competition among Thai inbound tour operators intensifies, the package prices charged by the inbound tour operators to their outbound Chinese counterparts starts to drop going to US$60, then US$40 and eventually reaches zero. This is the reason why this type of group tour is dubbed a "zero-dollar" tour. While this self-inflicted price war hurts many suppliers in the supply chain as the profit margins are squeezed out of them, it benefits some older hotels that have outdated facilities and are not attractive to more conventional group tour packages. The question is: how can these Thai tour operators and their suppliers benefit from this below-cost pricing practice? The reverse process then continues as follows.

Normally, tour guides are paid by the tour operator for guide services. In the case of the zero-dollar tour, tour guides pay the tour operators between 100 and 1,500 baht per visitor for the guide services they provide. Because tour guides (as suppliers) pay the operator to obtain customers, they use many tricks or scams to make a profit out of the visitors. Tour guiding services have now become selling activities; tour guides have become salespersons. Selling starts from the moment the visitors get on the tour bus. The originally announced schedule is discarded as the guides steer the visitors to shopping centers, jewellery shops, and even sex shows that pay them handsome commissions or kickbacks. Because transportation dominates a big portion of the tour operating costs, some jewellery shops even pay for renting the tour buses instead of letting the tour operators pay in exchange for tour guides bringing visitors to their shops.

If no one in the group tour is interested in buying goods from these contracted jewellery shops or shopping centers, the entire tour can come to a halt. The tour guides will not move on to the next program unless someone buys. Since the original schedule almost never takes place as planned, optional packages are then offered to tourists for an additional cost. This often results in constant bickering and quarrels between the tour guides and the members of the visiting group. Tourists are known to file complaints about the tour operators to the tourism authorities. However, no effective solution has yet been found.

The scams continue as a consequence of the collusion among the tour operators, tour guides, shops and other service providers. Visitors sometime end up paying extremely high prices for products/services. For example, the boat fare from the Thai border in Chiang Rai province to Burma/Myanmar is normally Bt 100. But some group tours charge Bt 500 instead leaving Bt 400 as a profit for the tour guides.

Results of a survey released by the Thai Chamber of Commerce University (TCCU) found that Chinese visitors pointed to cheap food and low-quality guides as their worse experience while visiting Thailand (China People's Daily 2002). Cheap food and low-quality guides are, in part, consequences of the low-budget and zero-dollar tours that offer substandard services.

Avoiding the "Low-Price Trap"

Most small tourism operators such as small transportation entrepreneurs feel that they are at the bottom of the supply chain and always taken advantage of in terms of the benefits distributed to them, In Chiang Mai area many locals are involved in transportation services for the tourism industry including rental van services. Rental vans are provided by both the kerbside van operators and individual van owners.

*Kerbside Van Network Operator*This is a micro-business with a fleet of vans parked and operating from the kerbside. It has no formal office; instead, a small table is used often located at the kerbside of major streets such as in Charoen Pratet Road near the Chiang Mai Night Bazaar. The network is formally registered as a van operator. The owner of the business does not own vans but uses a membership network of individual van operators. To become a member, an individual van owner pays a one-time membership fee of Bt 3,000 and 10% of the daily rental price. The daily rental price in 2007 was Bt 1,200 for an in-town excursion and Bt 1,500 for out-of- town trip. Code of conduct regarding driver behavior is strictly enforced. The members line up beside the road in the morning to pick up customers arranged by the network operator. Customers of the network are predominantly independent tourists both domestic and international. These tourists made their reservation in advance directly with the operator or at the hotel or guesthouse front desk.

*Individual Van Owners*Most individuals owning vans for rent used to work for tour operators and earned a monthly salary. After they learned some skills and the tourism channels, they quit as members and became independent (and unofficial) sub-contractors for tour operators. Unlike well-established transportation companies, independent van owners do not sign any contract with tour operators. There are three price ranges for van rental service: Bt 1,000–1,200 for a smaller and older van, Bt 1,200–1,500 for an older but wide van and Bt 1,500–1,800 for a new and roomy van. Most van operators realize that relying solely on tourism revenues would shrink their profits partly due to the high competition in the market. Therefore, most of them diversify revenues by servicing customers in non-tourism sectors including students and researchers who need transport for field visits. In this case, they can charge at the top of the price range, instead of earning at the low range by servicing visitors sent from tour operators.

The three main strategies used by individual van owners to obtain customers are: (1) good relationships with the tour operator (2) a network of friends in the van rental business and (3) providing service for non-tourism groups.

Although individual van owners are not officially associated with tour operators, they still rely on businesses supplied by the tour operators. The problem arising from receiving customers from tour operators is the "low-price trap". Usually, this offer price by tourist operators is lower than the market prices mentioned earlier due to high competition in the van rental market yielding very low profit margins to the individual van owners. The strategy of obtaining customers from their networks of friends complements the first strategy. The third strategy of servicing customers outside the tourism industry yields the highest profit as van owners are "price setters". It is also considered a "revenue diversification" strategy. Van owners have to search for customers by themselves. One way to obtain customers in this segment is to establish a good reputation among non-tourism-oriented organizations that need transportation service.

Although the Thai tourism market is highly competitive and profits are squeezed out of many suppliers, some firms still adopt sustainable business practices. For example, some tour operators have profit sharing plans to share the company profits with employees and local communities. A tour operator located in Chiang Mai pays a certain amount of money per visitor to local communities every time the company brings visitors to visit the villages. Some firms use quality of service to compete rather than lowering their prices. A locally owned car rental company refuses to rent cars to tour operators that offer too low a price. Instead, it adheres to its policy of offering high quality cars and services with the right price. The company's policy pays off handsomely. Currently, it has the dominant market share in the Chiang Mai area and it is expanding the business to cover many other northern provinces.

In the hotel sector, many hotels in Thailand pursue at least four sustainable business practices: employee benefits, reduction in water and energy use, recycle, facility inspection, and donation for attraction renovation.

Employment Most hotels employ local staff except for some positions that require specialty skills such as a chef for Japanese cuisine. It is common for most international hotel chains to employ foreign or non-Thai staff. However, some local hotel chains also have started to hire their managers from western countries. This may be due to the fact that managers from those countries know the customs and cultures of their main customers such as westerners better than the Thai counterparts.

Kranok Suvannavisutr, General Manager of The Empress, however asserts that his hotel has a policy of not hiring foreign workers. This is because Thais have a far superior hospitality attitude than their foreign counterparts.

Reduction in Water and Energy Use Kranok estimates that energy expenditure is approximately 50% of hotel operating costs. To save costs, all hotels have energy-saving measures. Many Thai hotels offer a choice for customers whether to have their bathroom towels changed daily. The following message is posted in the bathroom of some hotels "if you prefer to have your bathroom towels changed, please drop them on the floor. Otherwise, hang them on the rack." This message helps reduce the amount of water and detergent used for laundry. Some hotels such as Grand de Ville in Bangkok have the message posted in the bathroom reminding guests to help save water and energy. Needless to say, these practices reduce the operating costs of the hotel as well as create efficient use of water and energy.

Recycle In the hotel business, there are many bathroom supplies that can be recycled such as shampoo containers. Many customers take away with them tiny plastic-made shampoo bottles and, most of the time, those bottles end up in the garbage bin at home. To avoid this problem, the Tamarind Village Hotel in Chiang Mai uses big northern-style bottles made of glass as shampoo containers with the message hanging on the bottle neck "To preserve our environment this bottle is for refilling of daily bathroom supplies. For purchase, kindly contact our staff." Mid-price and cheaper hotels often use "dispensers" attached to the wall with liquid soap and shampoo.

Facility Inspection This is a self-regulated measure implemented by the Thai Hotel Association. One of the requirements for members of the Northern Chapter of the Thai Hotel Association is that member hotels must have a proper water treatment system. It can be noted that this practice does not apply to non-members of the Thai Hotel Association.

Donation for Renovation of Tourism Attraction Sites The profits of a hotel may depend on its location and the continued ability of nearby tourism sites to attract people. Thus some hotels provide financial support for renovation of tourism sites in the area.

Sanitation and Hygiene at Tourism Sites

One of the biggest concerns of tourists is illness (Reuters 2007). A survey by Tripadvisor.com, covering 2,500 visitors worldwide, found 80% of visitors were concerned about germs, bacteria and viruses when travelling. Nearly half of the travelers said that their worst experience while traveling is entering a dirty bathroom. The destination countries can minimize these concerns by providing adequate and good quality sanitation and hygiene services.

Service Provider The most common health related problems for travelers visiting developing countries is traveler's diarrhoea or TD. As many as 50% of travelers from industrial nations develop TD while visiting developing countries.

The US Center for Disease Control (CDC) attributes the causes of TD to eating contaminated food and water (Torres and Skillicorn 2004). Thailand is listed by the CDC as a TD-risk country.

From my observations at eight sites, one of the main weaknesses at tourism sites in Thailand is that services are frequently unsanitary and unhygienic. We will discuss three main services: toilet and bathrooms, vending and restaurants, and garbage collection.

Bathroom & Toilet Bathrooms at many tourism sites are not sanitary due to poor facilities, lack of amenities, inadequate maintenance, and poor behavior of visitors who use the bathroom. The high frequency of use and lack of adequate maintenance make some bathrooms, especially those at popular tourism sites in the area such as the Ayutthaya Historical Park give off a foul smell.

The behavior of visitors contributes substantially to the sanitation of the bathroom such as not flushing the toilet after use or spilling urine all over the toilet seat and on the floor.

It would be worthwhile to note that, from our observation, visitor behavior is also influenced by the existing condition and tidiness of bathroom facilities. We hardly saw a toilet left unflushed or urine spilt on toilet seats at restrooms that are extremely tidy, well maintained and provide enough amenities.

Vending & Restaurant Tourism attractions offer two types of vending: concession and mobile. Conventional mobile vendors sell preserved fruits, fast foods, bottled drinking water, ice cream, and other traditional Thai desserts. Concession vending areas sell souvenirs representing local uniqueness or identities and are usually organized and look cleaner and more sanitary than mobile vending areas. Improper hygiene is a problem with mobile vending who sell food as the food is not kept covered and remain exposed to dust and smoke from vehicles.

Food services at restaurants in tourism sites sell typical Thai fast food. Three issues that could pose health and hygiene threats to visitors are: flies, dust and poor facilities.

Garbage and Waste Disposal Garbage collection is problematic at many tourism sites, which could contribute to health problems for visitors. There are three main garbage issues experienced by tourism sites: low garbage carrying capacity, visitors' negligence, and limited space for garbage disposal. First, some tourism sites produce more garbage than local governments can collect. For example Rawai Cliff, one of the most popular spots in Phuket, hosts over a 1,000 of visitors per day coming to see the sun set. The influx of visitors means five medium-size dumpsters are quickly filled and sometimes overflows. The TAO has a garbage collection capacity of 25 tons per day but has to collect 30.

Second, despite tourism sites providing adequate number of garbage bins, those sites still experience a problem of garbage – mainly food and drink related items such as soft drink and alcohol bottles, food containers, etc. – being scattered all over the place or left negligently by visitors at many spots such as benches, picnic areas, beachside yards, beachfronts and parking lots.

The third issue is disposal. Some areas, especially small islands, lack suitable space to dump garbage. Yao Noi Island in Phang Nga Province in southern Thailand has experienced this chronic problem since the advent of enclave tourism resorts on the island. When the island is short of space for land-fill garbage disposal, garbage is taken and dumped in the surrounding mangrove areas.

In summary, sanitary and hygienic services offered at tourism sites are crucial to the sustainability of tourism. Unsanitary and unhygienic services are one of the main weaknesses we observed across a range of sites in Thailand.

Tourist Education and Culturally Acceptable Behavior

While tourism can benefit local communities, the behavior of some visitors may also cause cultural tensions and conflicts. A tourism attraction can be viewed as a temporary gathering place for people coming from different countries and cultures. Local residents often witness what they perceive as culturally inappropriate behavior of visitors. In order to reduce the negative effects of visitor behavior on local culture most major tourist attractions have declared a code of conduct (COC) to educate visitors.

The most common medium of communicating culturally accepted behaviour is using signs or bulletin boards (see Fig. 12.2). In addition, staff are employed at some attractions to deliver the COC messages. The purpose is to educate both the first-time and repeat visitors. However, not all visitors understand these messages. Some tourism sites display messages only in Thai language (Fig. 12.2a), and some display in Thai and English, leaving out many other major languages spoken by visitors (Fig. 12.2b). Most messages emphasize the conduct acceptable to the local cultures and about sanitation. For example, most temples request visitors to take off their shoes when entering the main Buddha hall while ecotourism sites request visitors to throw litter in the garbage bins.

There are various tactics used by tourism sites to draw the attention of visitors. For example, some sites use large signs in multiple languages to deliver their messages; some display the amount of fines that are imposed for violations; some write messages in Thai and ask the Thai people to remind international visitors about proper behavior; some signs use indirect and sarcastic language (Fig.12.2c); and some use universal symbols (Fig. 12.2d). Various types of signs have different effects on visitors.

Despite the signs, it can be observed that violations are still prevalent. This raises two main issues: the effectiveness of signs and improper visitor behavior. The first issue is tourism sites may not properly display the message. Tourism site managers should test the effectiveness of COC signs and bulletins. The second issue is that many visitors are ignorant, stubborn and arrogant and may seek to disregard or violate the codes regardless.

Effectiveness of Signs In general, signs displayed by tourism sites are relatively effective. But it is not always easy for foreigners to follow the instructions for example when signs are written only in Thai language or not posted on the right spots. For example the Phuket Galleria, which has three floors, posts the following signs only on the second and third floors "Please do not take a photo or touch

Fig. 12.2 Some information signs. (**a**) Sign in Thai asking Thai visitors to remind the international visitors to not violate the code of conduct, (**b**) an eye-catching red sign with a message about proper attire displayed in front of the Buddha Image Hall at Wat Chalong in Phuket, (**c**) a Thai sign is aimed jokingly for monkeys: Announcement from the Guardian Spirit of Khao Yai – Monkeys are strictly not allowed to receive food from humans. Violation shall bring grave danger to both parties (monkeys and humans). If you (monkeys) don't believe in these messages, don't act disrespectfully. May I remind you that many of you, including parents and children alike, have been hit and killed by passing vehicles, (**d**) to avoid language barriers, some signs use international symbols

pictures and sculptures." But since the message is not posted on the first floor, visitors may unknowingly violate the gallery's COC. When we asked the gallery attendant, we were told that taking picture is prohibited everywhere in the galleria including the first floor.

Ignorance Despite tourism sites having effectively declared COC messages, using information persons or security guards, some visitors may still violate COC. This may be because some visitors are ignorant, stubborn or arrogant or don't like being told what they should not do. However, sometimes it may also be a need for modifying the rules.

The Phuket Aquarium and Ayutthaya Historical Heritage Sites are good examples of observable violations.

The Phuket Aquarium is a privately owned attraction and one of the most popular spots in Phuket hosting thousands of visitors per day. The aquarium has the signs "Please do not use flash when taking pictures." and "Do not touch the aquarium" posted in front of every aquarium tank. However, approximately 40% of visitors both international and domestic used flash and touch the aquariums.

At Ayutthaya Historical Site, a world heritage site in central Thailand, all signs are written in Thai, English and Japanese, which read "Please don't climb on the stupa", "Please do not take photographs of yourself with the head on the top of a headless Buddha statue" and "Do not climb the wall". Our observation indicates that approximately 10% of visitors, mostly foreigners, disobeyed the signs and performed the prohibited actions. The Department of Fine Arts (DFA) hires security guards to watch out for visitor behavior and "blow their whistle" to remind them that they are breaking the rules. Guards then approach violators and inform them that what they are doing is breaking the code of conduct. Breaking local rules have detrimental consequence as a villager residing in the heritage area put it: "When we see visitors climbing the stupa or Buddha statue, we feel that these people show no respect to our culture."

It is recommended that tourism sites that experience high rates of violation should consider testing the effectiveness of signs. In addition, signs should be convincing enough to alter visitors' behavior. This may be implemented by posting the reasons why some actions are prohibited. For example, "Please do not climb the stupa because doing so is culturally inappropriate and may also cause the stupa to collapse."

It could be noticed that the patterns of breaking the visiting rules differs from group to group of visitors. Thai visitors are notorious for littering or feeding animals while international visitors are known for having manners that are culturally inappropriate. Among international visitors, the behavior differs by the type of tour and the place they visit.

Type of Tour and Tourist Behavior There are mainly two types of tours: group tour and independent traveling. Group tour members are likely to have less sustainable behavior toward local cultures than independent travelers. Group tour members may not pay full attention to COC messages delivered by tour guides. In general, group tour members travel in big groups. From our observation, some activities such as conversation with group mates distract group tour members from paying attention to COC messages. In addition, group tour members are likely to expect local information to be fed by tour guides. What if the tour guides fail to deliver COC?

A tour-guide-turned- businessman admits that, based on his experience, it is possible that tour guides fail to deliver or explain COC because they are always preoccupied with many issues that require spontaneous responses and actions such as paying entrance fees, monitoring touring schedules and answering the questions of group members. Consequently, group tour members may not be fully aware of COC.

Group tour members may not seek local cultural information beforehand. Instead they rely on information to be fed by tour guides as they visit tourism sites. When tour guides fail to deliver the message, tourists may violate COC unknowingly. On the contrary, my observation and interviews with some FITs shows that foreign independent travelers (FITs) are more likely to obtain cultural information in advance, making them fully aware of the local COC. Most of them travel without tour guides; they rely on a pocket guidebook and learn about the local culture before arriving at the destination.

Many Western women wear short pants with tight and/or sleeveless shirts while traveling in Thailand. At Chalong Temple in Phuket, the author observed that many of them had brought their own sweater or traditional Thai cotton skirts which they changed into before entering the site. At Phrasingh Temple in Chiang Mai, I saw an American couple entering the site without reading the COC signs. Before entering the Buddha Image Hall, the wife, who was wearing a tight sleeveless shirt stopped right before the steps to put on the sweater brought with her in her backpack. When I asked her, she told me that she needed to wear the proper mode of dress before entering the Buddha Image Hall from a guide book.

Places and Tourist Behavior Regarding tourist education via signs and bulletin boards, cultural attractions tend to educate tourists better than sea-sand-sun (3S) tourism. We observed how tourist attractions educated visitors and how visitors responded in two distinct tourism activities: cultural and ecotourism sites in northern Thailand and 3S tourism in the south. While most northern provinces offer both culture tours and ecotourism, most southern provinces including the most popular spot Phuket represent 3S tourism. From our observation, visitors who visit cultural and ecotourism areas tend to behave more sustainably toward local cultures than those visiting 3S tourism. For example, walking topless on the street in Thailand is viewed by Thai people as culturally inappropriate behavior. We hardly saw a male visitor walking topless on the street of Chiang Mai, a culture tourism city, but we saw a lot of them on the streets of the beach city of Patong in Phuket. Also in the City of Pattaya, one can frequently see bare-chested male visitors walking on the street (Cleary 2006). The visitors behaving differently in different settings may imply the following:

Firstly, the types of tourism shape visitors' awareness of sustainability. Ecotourism and cultural tourism attractions offer natural and cultural features that easily convince visitors to abide by local norms and traditions. On the contrary, 3S and mass tourism offer no cultural uniqueness, making visitors unaware of local norms and traditions.

Secondly, tourist education influences visitors' awareness. Our observation indicates that ecotourism and cultural tourism attractions educate tourists, using signs/bulletin boards, better than 3S tourism attractions. We hardly saw signs educating tourists at 3S tourism areas we visited.

Lastly, the image of a place attracts different types of visitors who behave accordingly. Phuket not only has beautiful beaches, but it also has images of sex tourism, drugs and night-life entertainment.

In summary, tourism sites educate visitors by delivering culturally acceptable behavior messages posted on signs/bulletin boards. The majority of visitors behave accordingly. Some sites experience high rates of violation; consequently, they employ information persons and security guards to watch out for violations. Visitors who visit cultural and ecotourism attractions are likely to behave better toward local cultures than those visiting 3S tourism. FITs tend to conform to local cultures than their group-tour counterparts.

Enhancing Sustainable Tourism: A Public Policy Perspective

Tourism is unlikely to be sustainable without the involvement of the government. From a policy perspective, there are many challenges that governments will face to create sustainability in the tourism production and consumption processes. Those main challenges are: distributional goals, externalities, public goods, and monopoly of services.

Distributional Goals and Local Income Distribution Schemes

When policy-makers talk about tourism, they often mention the number of visitor arrivals and the financial benefits for the local economy. Not many (or probably "none") mention financial losses (leakages) and how tourism benefits are distributed among various groups involved in tourism. The distributional issue may not receive much attention from policy makers. However, the issue is widely discussed among villagers: the more tourism is developed, the less are the benefits they receive. Three issues concerning distributional goals are discussed here.

First, tourism distributes benefits to local residents, but some may not be aware of it. Thongchai Wongrienthong and Udom Chidnayi; the deputy governor of Chiang Mai province and the owner of a tour company respectively, gave the same example: when tourists dine in local restaurants or hotels, they distribute income to farmers who grow rice and vegetables. In other words, local farmers supply ingredients for the production of finished goods to be consumed by tourists. In addition, tourism-related taxes collected by local governments are distributed to locals in the form of infrastructure improvement. Both examples are correct. In fact, not all local residents directly participate in the tourism supply chain. In other words, not all residents have careers that are related to providing goods and services consumed by tourists. However, those who do not directly participate in the supply chain become indirectly involved in tourism production because they are subjects to be viewed for the pleasure of tourists. Therefore, people in local communities should be fairly compensated.

Second, have villagers done enough to benefit from tourism? Many people believe that tourists bring money to the local community. Some argue that this is not necessarily true unless locals make them spend. How much money is spent in the local economy and how effective is the income distribution is partly determined by the extent to which villagers invest their effort and resources to attract tourists to their community. According to a district chief in Chiang Mai province, villagers have to know how to take the money out of visitors' pockets. He refers to products, designs and packaging of local souvenirs and other services that must be attractive enough to encourage visitors to buy or use local products/services.

In some districts, local committees or working groups manage and find measures to make tourism spending benefit local people and to distribute tourist spending equitably. For instance, residents of Phi Mai District and the Kao Yai area in Nakorn Ratchasima province form housewife groups. Each group possesses different skills and produces different products accordingly. Households become members of a group on a voluntary basis. Income derived from selling product/services such as massage, home-made Thai silk, and food are distributed among members. In some community-based tourism areas, tourist spending is not only allocated to member households, but it also given to village funds for the purpose of community development.

Third, have governments, local and national alike, done enough to distribute benefits equitably across various groups? According to the Thai Tourism Satellite Accounts produced by World Travel and Tourism Council, between 1998 and 2005, international tourism accounted for 13% of GDP (Bt 655 billion), 10% of employment (3 million jobs), 13% of exports (Bt 417 billion), 12% of investments (Bt 117 billion) and 3% of the government budget (Bt 13 billion) (Wattanakuljaras 2006).

Anan Wattanakuljaras (2006) found that tourism growth benefited all household classes in terms of an increase in overall consumption, income and welfare. However, much fewer benefits were distributed to the poor. This coincides with the notion from villagers in many areas that the more the tourism is developed, the more benefits fall to big business leaving very little to locals. It is true that benefits to local residents diminish when tourism moves to higher stages of development (Butler 1980). However, governments could intervene to counter the cycle by using income distribution schemes.

One of the most effective income distribution schemes is to have local residents take direct part in the tourism supply chain. Krata-Kraron Beach Municipality in Phuket is successfully implementing this scheme. Villagers are organized into 19 tourism career groups based on their expertise: for instance, long-tailed boat tour, beach umbrella services, food and beverage services and local transportation groups. These groups and their members have to be registered with the city. The groups work closely with tourism associations in the areas. Hotels supply tourists to these local career groups. According to the City Clerk, the success of this scheme is attributable to strong local government involvement. The government acts as coordinator and facilitator between tourism associations and villagers' career groups. In addition, government intervention prevents the problem of "mafia" groups that can erode the income of the career group participants.

I suggest another scheme that could be implemented by Provincial Administration Organizations (PAOs), the highest level of local governments. Currently, some PAOs subsidize the cost of overseas road shows for some hotel associations based on the principle that some tourism-related taxes are collected directly from hotels.

My view is that hotel guests not only stay in the hotel; they also take excursions to local destinations in the areas. Therefore, local residents living in the tourism areas should be paid as well for their effort in providing and preserving tourism resources. Payments can be made either in-kind or in-cash, such as financing pro-poor projects in the tourism areas, renovating tourism attractions, or allocating money to village development funds.

Finding Effective Ways to Deal with the Problems of Negative Externalities

Externality is a valued impact (positive or negative) resulting from any action that affects someone who did not fully consent to it through participation in voluntary exchange. Negative externalities in the Thai tourism production system are caused mainly by producers. Water pollution as a result of irresponsible business behavior is a common example.

Residents of Koh Chang Island, located in the eastern Province of Trat, had suffered from the unpleasant smell of waste water released to the sea by businesses in the area. The Koh Chang TAO has no water treatment facilities. It allowed tourism operators such as resorts and restaurants in the area to release waste water without proper treatment into a small stream that leads to the sea. Villagers and vendors had been patient for 2 years. Because of foul smell and loss of customers visiting their shops, they decided to call in a national newspaper to investigate the issue. Waste water from the stream was released to the sea at the famous Had Sai Khao Beach turning the sea water at the beach into black (The Manager Online 2008a).

A second example is the case of villagers residing in Moo Si village located at the Kao Yai foothill of Kao Yai National Park who reported foul smell coming from untreated waste water released by a hotel into a stream. Residents of Pattaya and Krata Kraron municipalities filed complaints against a hotel releasing untreated water into the ocean and stench.

Everything necessary to solve these sort of problems is already incorporated in existing environmental and public health laws, but there is little enforcement: if you know people in power, you can get away with nearly anything.

Filing Complaints and Lawsuits: An Effective Way Out

Filing complaints may be considered the most effective way to enforce laws in Thailand. In general, pollution released from tourism businesses has two simultaneous effects: it pollutes the environment and releases an unpleasant smell to adjacent communities. This gives an opportunity for residents in the community to use their property rights to file complaints against polluters.

Villagers at Banglamung District in Pattaya area filed a complaint with Pattaya City authorities against a hotel releasing untreated waste water into the sea in the middle of the night, causing foul smell and water pollution. The Waste Water Inspection Department assigned a team to investigate the issue and found the hotel guilty of discharging waste water into the sea without proper treatment. The department ordered the hotel to clean up and improve water treatment facilities and not to release waste water into the sea without being treated properly. However, the hotel ignored the order. A few months later, the authorities called in the hotel representatives to inform them that a second complaint had been filed by villagers due to the hotel's non- compliance. The hotel representatives refused to negotiate and walked out of the meeting. Finally, the City of Pattaya decided to file a lawsuit against the hotel at the Provincial Court. The Director of the Waste Water Inspection Department said "In fact, the City of Pattaya did not intend to be stringent in enforcing the law. Because the department received frequent complaints against the hotel, a team was set up to investigate the case" (The Manager Online 2008b).

Monopoly in Service and Substandard Quality

Many tourism sites provide sub-standard facilities and services, for instance unsanitary food and dirty bathrooms. This is partly due to the fact that services provided by a tourism site are monopolized by a tourist operator. This happens mainly with small and medium tourism attractions run by local governments, where only one or two service providers are contracted. There may be, for example, only one concession restroom facility and one or two restaurants.

In contrast, big tourism sites such as privately owned fish farms in Suphan Buri province and most national parks run by the National Park Services (NPS) and the Department of Special Economic Zones offer a wide variety of services for visitors to choose from. There are many concession vendors and restaurants in the premises and many restroom facilities that are well maintained. Hence, food and vending services and restroom facilities provided at those sites are likely to be more hygienic and sanitary than those operated at small tourism sites.

We observed that visitors were more likely to use the service of tidy restrooms than those with substandard services, despite both charging the same price. The crucial factor determining higher consumption is service quality rather than user fees.

Although it may appear amusing to recommend that tourism attractions use high-standard restrooms together with clean food service as selling points to induce more tourist spending, in fact, using clean restrooms as a marketing tool has proven effective for years by some gas stations in Thailand. The Thai Gas Station Law requires that gas stations provide restrooms for customers. Since the quality and price of gasoline sold at each station are nearly homogeneous, many gas stations have put up a conspicuous sign "Clean Restrooms" to attract customers. Some gas stations also obtain a "Clean Food Good Taste" certification from the Public Health Ministry.

To elevate the standard of food and restroom services at all tourism sites in Thailand, a third-party certification may be used. The third-party certification, such as

ISO and GMP, has been widely used by big businesses. However, it is rarely used in tourism site operations. Currently, the Ministry of Tourism (MoT) has implemented home-stay standards. Home-Stay Certification has been awarded to qualified communities that opened their villages as accommodation for visitors. The MOT should also set up and implement the bathroom standards for all tourism sites. The standards should cover among other things: facilities, frequent maintenance, and amenity requirements. Standards of garbage collection and disposal should also be set.

In addition, the ministry should enforce food service standards. Currently, the Public Health Ministry awards "Clean Food Good Taste" to qualified restaurants. Some local governments are taking their own initiative to implement food standards. For example, Bang Saen Beach Municipality, 30 miles east of Bangkok is implementing food standards in its jurisdiction by asking food vendors in the beach area to obtain "Clean Food Good Taste" certification from the Public Health Ministry. This initiative should be adopted at tourism sites across the country.

Using Public Good Principles to Enhance Sustainability

Public good is a commodity or service whose consumption by one person does not preclude others from consuming it as well. Public goods are non-rivalrous in consumption and non-excludable in use (Weimer and Vining 1999). A good that has both characteristics is pure public good, while a good that has either characteristic is quasi-public good. A common problem arising from consumption of public good is overuse. To prevent overuse, many measures can be imposed such as quotas and user fees.

Tourism attractions can be classified as quasi-public goods, which have three main policy implications to maintain sustainability: user fees, excludability, and carrying capacity.

User Fees Many tourism sites collect user fees, which are based on price discrimination principles. In general, international visitors pay higher user fees than domestic visitors. This practice is common among tourism attractions in developing countries. There are some issues regarding collecting user fees at tourism sites in Thailand.

First, how should user fees be allocated according to the principle of quasi-public goods? Many tourism sites in Thailand provided sub-standard services and facilities such as unsanitary bathroom and restaurants despite collecting entrance fees. To raise the standard of services, a portion of entrance fees should be earmarked for the costs of maintenance. To put it differently, part of those fees should be used to increase garbage collection capacity, to provide necessary amenities and proper maintenance of restrooms, and to provide higher standards of food services. Because tourists pay to use tourism services, they should receive high service quality in return.

Second, some tourism sites do not collect user fees and provide sub-standard services. For example, Rawai Cliff, one of the most popular tourism attractions in Southern Thailand, attracts thousands of visitors every day. But the restroom and garbage collection services are of poor quality. Collecting a small amount of entrance fees, such as Bt 10 ($0.3), to finance the operation could enhance better service quality.

Third, user fees collected from centrally-managed tourism sites located in local government jurisdictions should partly be transferred to local governments. This is because local governments provide basic infrastructure to tourists visiting those sites as well as to the operation of the sites such as garbage collection service and local roads.

Excludability Another issue regarding the public good nature of tourism is excludability. No one would be excluded from consumption of public goods unless the goods themselves are facing a congestion problem. Evidence from interviews reveals that villagers residing in some famous beach areas were denied access to the beach citing the reason that beachfronts were reserved for hotel's customers. In fact, all beachfronts are public areas (public goods). Hence, no one should be denied access. How can tourism be sustainable if local residents feel that they are inferior to hotel guests and excluded from using public goods in their own hometown? Those excluded can be considered silent losers, who receive high risks associated with tourism development.

Carrying Capacity Tourism as a quasi-public good has a common problem of congestion mentioned previously. In other words, although consumption of the good is not excludable, not all can consume at once due to the limitation in availability of the good, such as space and services. This implies that carrying capacity has to be taken into account when receiving visitors. This issue has been studied extensively and is one of the most popular issues in tourism in Thailand. Therefore details will not be discussed here. However, there are two points worth mentioning: the attitude of "the more, the better" and spending on advertising and marketing supersedes maintenance. Policy makers tend to focus more on increasing the target number of international arrivals than to pay attention to maintaining or improving the carrying capacity of tourism attractions. This is manifested in the government allocation of budget for tourism promotion. The advertising budget is an annual budget but the maintenance budget is a special budget that needs the discretion of the Council of Ministers despite the fact that tourism sites suffer degradation every year. The marketing budget occupies the lion share of 70% of the tourism related allocations for the national tourism authorities.

Limits to Public Intervention

Tourism in Thailand is not likely to be sustainable without government intervention. However, government intervention in the market itself can obstruct sustainability. This is because policy implementation has some limitations as follows.

Rent Seeking and Special Interests

Tourism policy in Thailand is a "silent policy" because there is no public participation in the policy making process. In general, private self-interest plays an important role in motivating public participation. In Thailand, policy debates are prevalent in selected areas such as economic policy, public health scheme and village funds. Although tourism is the fastest growing industry in Thailand that accounts for more than 13% of GDP, tourism policy is rarely debated publicly. The tourism policy direction had been predominantly set by bureaucrats in the Tourism Authority of Thailand (TAT). The lack of public participation in policy formulation leaves the tourism policy to be influenced by vested interests, such as big tourism businesses. This is contrary to, for example, tourism policies in most Caribbean countries, which draw a lot of public attention and policy debates (Shamsub et al. 2006).

Due to the sharp decline in the number of tourist arrivals in 2001, the then government of Prime Minister Shinawatra decided to make tourism policy the main engine of economic growth.

Tourism policy in Thailand concentrates on benefits while ignoring the costs to society. The focus is on setting high targets for the number of visitors. This puts pressures on the tourism marketing agencies to meet targets by opening the door for low quality tourist packages involving sex and drug tourists and zero-dollar tours. Powerful vested interests are behind these tourism ventures making these ventures hard to eliminate.

In addition, government interventions results in rent seeking. For example, the effort to rejuvenate tourism in Chiang Mai by the government was severely criticized for rampant rent-seeking practices in the operation of the Chiang Mai Night Safari (The Nation 2006).

At a local level, many TAOs rely heavily on revenue from tourism, but they have no tourism plans. Many TAOs we interviewed insinuated that this was because local policies were dominated by vested interests behind local governments. Some community leaders prefer to implement utility and infrastructure projects because those projects require construction, which could easily yield "rents" to people in power and their cliques. Local tourism policies normally do not receive much attention from community leaders because it does not involve many large construction projects with budgets for construction contracts.

Lobbying is prevalent at the local level. Environmental laws require water treatment facilities at hotels be inspected periodically. Some residents including local politicians we interviewed pointed out that, in practice, facilities in many hotels were hardly inspected due to the close relationships between hotel businesses and environmental officials.

Central Agencies vs. Local Governments

Most centrally-owned tourism attractions are located in local government jurisdictions. However, local governments have no legal rights to share management responsibilities and user fees like entrance fees, although they provide basic services such as sanitation and garbage collection. The increased number of visitors in the area leaves many problems to be solved by local governments; for instance, a surge in the amount of garbage, local road surface deterioration and traffic jams. But local governments are not financially compensated. For some local jurisdictions, the costs of carrying tourists visiting centrally-managed attractions may be greater than benefits received by local governments.

This problem is exacerbated in the jurisdictions that visitors pass through for an hour or two to visit centrally-managed sites, not final destinations. Examples include the City of Phi Mai and Ayutthaya City. Visitors normally visit Phi Mai Castle and Ayutthaya Historical Park for an hour or two per day; then return for overnight accommodation in Nakorn Ratchasima City and Bangkok respectively. Visitors visiting those sites hardly spend anything locally. Most items they spend their money on are entrance fees and a bottle of water. In other words, they don't dine locally or stay in local accommodation; consequently, little money goes into the economy of local jurisdictions. However, their presence for a short period of time may incur more costs than benefits to local governments. One way to solve this problem is to have local governments share management responsibility with centrally-managed sites. Another is to have a revenue sharing plan between central government departments running the sites and local governments.

Local Participation

Many tourism projects were implemented without consultation with local residents. This causes resentments among locals. Some development projects have been cancelled and delayed for years due to protests. The Doi Pui Development Project in Mae Hong Son Province, proposed by Mae Surin Waterfall National Park Service in 2005, was cancelled abruptly amid a series of protests by villagers. The plan to expand Phi Mai Historical Park in Nakorn Ratchasima province has been put on hold for over a decade after it faced fierce resistance from villagers afraid of losing land through state expropriation.

Conclusion: Knowledge to Policy Recommendations

Sustainable Business Practices

Tourism development in Thailand as elsewhere is still far from being sustainable. Nevertheless in the tourism business sector, some sustainable business practices are evident as follows:

Some businesses are competing using high quality products instead of competing on the basis of price. In the highly competitive market, instead of using a below-cost or low profit margin pricing, a local car company, "North Wheel Rent A Car", adheres to its strategy of providing superior products and service quality with the right price. The strategy helps the firm to become a market leader in the area. In the hotel industry, many hotels post signs in guest rooms asking visitors to help save water and energy. Some signs ask guests not to take away glass-made containers because they will be reused for refilling liquid soap, shampoo, and conditioners

Some hotels provide cultural information regarding behavior that is perceived by locals as culturally inappropriate. A tour operator uses a benefit-sharing plan, which splits profits between local residents, employees and the firm. In addition, to help sustain tourism in the area, some companies make donations to renovate historical sites.

In the tourist attraction sector, a sustainable practice is to use various techniques to educate visitors. In many signs, the way of writing, the language used and their positioning can be very effective in changing visitor behavior. To increase effectiveness, some sites use information staff and security guards in addition to signs and bulletin boards. In addition, many natural attractions deliver environmental and ecological information. However, the practice of delivering code of conduct is still not adopted sector-wide. While most cultural and natural attractions adopt this practice, 3S tourist attractions are still lagging behind.

Government Intervention

To encourage all stakeholders to pursue sustainable practices in Thailand, as a country that has weak law enforcement and low institutional effectiveness, strong policy intervention from the central government is needed. To do so, the national government will have to face many policy challenges that address the following issues: income distribution, pollution, cross-cultural tensions, and sanitary and hygienic conditions of services provided.

Income Distribution The government has to acknowledge the reality that as tourism has moved to higher stages of development, local residents tend to receive less financial benefits from it. Therefore, the government has to counter the cycle by having local residents take direct part in the tourism supply chain. A good practice is to encourage local residents to form tourism-related careers groups such as beach umbrella services, local transportation, and traditional massage groups. Local governments act as coordinators and facilitators between these groups and tourism businesses as effectively done by the Krata-Kraron Beach Municipality.

Pollution There are many parties responsible for the problem of pollution such as firms releasing untreated waste water, local governments lacking water treatment capacity, and relevant authorities not enforcing laws. People who receive direct negative impacts from pollution are residents in the community. One effective way to reduce pollution is to encourage residents in the community to use their property rights and file complaints against polluters with relevant authorities. This strategy met success in Pattaya and Koh Chang Island. Reducing pollution is tantamount to enhancing environmental sustainability.

Cross-Cultural Tensions To minimize the chance of local residents seeing culturally inappropriate behavior of visitors from cultures that differ from their own, all tourism sites, including 3S attractions, should display cultural information and code of conduct on signs or bulletin boards. The effectiveness of signs and information should be tested. In addition, cultural changes in tourism communities should be monitored periodically.

Sanitary and Hygienic Conditions of Services One of the main weaknesses in the Thai tourism industry is the provision of substandard services in terms of hygiene and sanitation. Two possible practices that can lift the quality and standards of services are: A portion of entrance fees at tourism sites should be earmarked for maintenance of facilities such as restrooms, food services, and garbage collection; national tourism authorities should create a third party to set and enforce standards for tourism services. For example, MOT sets and enforces standards for restrooms and garbage collection. The Clean Food Good Taste certification from the Public Health Ministry should be enforced for food services at all tourism sites. This certification is being implemented at Bang Saen Beach Municipality. Sanitary and hygienic services reduce health risk to visitors and eventually will increase income to the community.

Making Tourism Policy Work Effectively: Knowledge to Action

The tourism industry consists of many sub-sectors and many stakeholders. The current tourism policy is formulated principally by the government tourism agencies and big businesses. This is a reason why tourism policy is mainly focusing on constantly increasing the target number of visitors and opening new attractions, rather than on addressing the negative dimension such as pollution, worsening impacts on local incomes, and declining service quality. To make sure that tourism benefits are fairly distributed across stakeholders, especially under-represented ones, tourism policy should be formed and administered by a national committee, such as the National Tourism Policy Committee, that encompasses representatives of various relevant bodies. Those include the national tourism marketing agency; tourism development agency; representatives from tourism businesses such as hotels, tour operators, small businesses; representatives from government agencies such as the Ministry of Education, the Ministry of Public Health, the Ministry of Culture, Police Department and local governments; and representatives of civic groups. While the tourism development agency sets tourism standards, civic groups ensure citizen participation. The committee should take charge of setting a comprehensive tourism strategies and policies that include tourism master plans and sustainable development strategies. In addition, the committee places emphasis on enhancing cooperation among stakeholders. This should resolve the problems of conflicts between local governments and national tourism agencies and ensure smooth cooperation between the public and private sector.

From the production perspective, tourism services providers should incorporate sustainable development principles in their corporate strategy. Under the sustainable development strategy, some simple measures could be implemented such as having a benefit sharing plan, competing using product or service quality instead of self-inflicted low-pricing; making donation to renovate tourism sites and reducing waste in the production process. For tourism development agencies, involving local resident and local governments in the tourism development process would be considered a good conflict resolution measure. Benefit sharing plan between tourism development agencies and local governments will ensure the right level of public goods provision in order to support local tourism development and to enhance better standard of living of local residents. For tourism sites that experience high level of violation in code of conduct, effectiveness of signs/bulletin boards for visitor education should be tested. Messages should be convincing enough to influence visitors' behavior.

From the consumption perspective, tourist behavior contributes greatly to tourism sustainability. Unsustainable tourist behavior could be detrimental to the tourism industry in the long run. Ideally, national tourism marketing should target tourists who have sustainable behavior, such as those who are environmentally and culturally conscious and those with positive contribution to the local economy. Because FITs are likely to have more sustainable behavior compared to group tours, more resources and effort should be allocated to draw this group of visitors. However, it would not be realistic to have only visitors with high level of sustainable behavior to visit the country. Due to the fact that visitors have mixed behavior, tourist education programs are needed. It is the responsibility of tourism service providers to ensure tourists' behavior conforms with local codes of conduct and to be environmentally conscious.

In addition, the image of a place could contribute greatly to tourist behavior. For example, some famous beaches have negative images such as child prostitution. For the negative images to be reversed requires the strong determination of the government to monitor these tourism areas and curtail negative activities.

In summary, to enhance sustainability in the tourism production and consumption systems, all actors have to work in concert. The central government has to take the initiative by framing policy guidelines for tourism sustainability. Policy execution should be supervised by the National Tourism Committee consisting of representatives of all tourism stake holders including government, business, and the civic sectors. Tourism policy should focus not only on benefit distribution but also risk reduction. Fair distribution of benefits and reduction of risks exposed to all stakeholders are essential to avoid conflicts among stakeholders that could hinder the progress towards sustainability.

References

Butler RW (1980) The concept of a tourist area cycle of evolution: implications for management of resources. Can Geogr 24:5–12

China People's Daily (2002) Bad food, poor guides hurt Thai image: Chinese tourists. Beijing, China

Cleary S (2006) Let's continue with the crackdowns. The Nation, 7 October 2006, Nation Multimedia, Bangkok
Curtin S, Busty G (1999) Sustainable destination development: the tour operator perspective. Int J Tour Res 1(2):135–147
Manager Online (2008) Fears of losing visitors: Businesses on Koh Chang ask authorities to solve garbage and waste water problems (in Thai). The Manager Online, 8 March 2008, The Manager Group, Bangkok
Manager Online (2008) City of Pattaya prepares to file a lawsuit against a hotel releasing untreated waste water to the sea (in Thai). The Manager Online, 8 March 2008, Manager Group, Bangkok
McKercher B (2003) Sustainable tourism development-guiding principles for planning and management. In: National seminar on sustainable tourism development, Bishkek, Kyrgystan
Reuters (2007) Germs top list of vacation worries. http://www.canada.com/com[onents/print.aspx?id=dfbc9e7d-d3ff-45d2. Cited 15 July 2008
Shamsub H, Albrecht W, Dawkins R (2006) Relationship between Cruise-ship tourism and stay-over tourism: a case study of the shift in the Cayman islands' tourism strategy. Tour Anal 11(2):1–10
Sinclair MT (1998) Tourism and economic development: a survey. J Dev Stud 34(5):p1
Tangwisutijit N (2004) Battered beaches. The Nation, 7 November 2004, Nation Multimedia, Bangkok
Torres R, Skillicorn P (2004) Montezuma's revenge: how sanitation concerns may injure Mexico's tourist industry. Cornell Hotel Restaurant Adm Q 45(2):132–144
The Nation (2006) Jaruvan orders probe into ambulance purchase, sets sights on Night Safari. The Nation, 28 October 2006, Nation Multimedia, Bangkok
Wattanakuljaras A (2006) 'Unseen' Thai tourism: where does the bath go?. Bangkok Post, Bangkok
Weimer DL, Vining AR (1999) Policy analysis: concepts and practices, 3rd edn. Prentice Hall, New Jersey, p 75

Chapter 13
Tourism Products, Local Host Communities and Ecosystems in Goa, India

Ligia Noronha

Introduction

This paper seeks to explore sustainable production–consumption systems (PCS) in tourism by examining the case of coastal tourism in Goa, India. We argue that sustainable production and consumption in tourism may be understood as tourism that respects the well-being of host communities and ecosystem health[1] in the tourist destination, more specifically the concerns over land and water while meeting the requirements and expectations of the tourist as well as the economic imperatives of the industry. Tourists are not a homogenous community. They have differing needs, tastes and spending power. Tourism-related infrastructure either reflects this diversity or through marketing influences the type of tourist that visits a destination. By sustainable PCS, we refer not only to the existence of environmental sustainability but also social sustainability. Sustainable PCS in tourism thus needs to ensure greater "tourist experience" for less material usage and host stress.

Tourism actually involves three integrated components on the production side: accommodation, transportation, and equipment and services for leisure activities; and it involves stakeholder groups at three levels: local, national and international. On the consumption side, it involves people with consumption needs, both basic and those created by the tourism industry. Moreover, there is a need to explicitly recognize the role of local host communities as both producers and consumers of the tourism product, and to understand this endogenous variable that lends its own dynamic to the social and environmental transformations that the tourism product generates. Host well-being is an important aspect of both the production and

L. Noronha (✉)
Resources and Global Security Division, Tata Energy Research Institute (TERI), New Delhi, India
e-mail: ligian@teri.res.in

[1]By ecosystem health we mean their ability to provide humans with the goods and services that are required for their continued well-being over time.

L. Lebel et al. (eds.), *Sustainable Production Consumption Systems: Knowledge, Engagement and Practice*, DOI 10.1007/978-90-481-3090-0_13,

consumption subsystem. It is not possible to disaggregate as the hosts are part of the production and consumption subsystem.

Discussions on the sustainability of production and consumption systems thus need to be located in the context of destinations, as most tourism related environmental and social impacts are local and contextual. Tourist destinations have a life cycle, from creation, through maturation, obsolescence and decline. Tourism production choices and consumer practices, in interaction with the local context, determine the shape and duration of this life cycle. Tourist destinations need, however, to be seen as part of larger systems, "nested hierarchies", as tourists, by definition come from other parts, national or international, with their own consumer needs and preferences; tourist practices and producer choices respond to both existing and new consumer preferences.

The product, both in terms of quantity and the type of resources that it utilizes and the changed relation that it generates in host destinations, creates impacts and brings changes in the social structure of local communities, the norms and standards of the host communities, and the environmental resources that it utilizes (TERI 2000). Because it has a spatial, place element, tourism activities interact with landscapes, occupations and ecosystems to alter these in favour of tourism. It is the *intensity* of the impacts rather than their *nature* that alters from one destination to the next, according to a range of characteristics that set the context of tourism development and help to determine whether or not it is a 'good thing'. The impacts that it has on the ecosystems and the social structure require that we question whether current tourism practices and models would enable the product to be consumed and produced without stressing these beyond repair, and whether the sustainability of tourism in a particular destination is intimately bound with the impacts on consumption that production choices create.

Very often notions of sustainable production and consumption are linked with notions of carrying capacity of a place. Calls are often made to restrict tourism in places where it degrades or threatens to cross thresholds at which the levels of stress will lead to the disruption of the ecosystems. Assuming that it is feasible to identify scientifically sound carrying capacity figures and that tourism flows are actually kept within the established carrying capacity limit, the question could still be asked: how useful is this concept in limiting consumption and production or the "tourist product" to the destination's ability to absorb impacts. Is it the most limiting factor that determines the 'true' carrying capacity which may not necessarily be related to the ecosystem capacity? A destination may receive fewer tourists than an environment can support, but more than its local population will accept. On the other hand, perceptions may change over time. Additional tourists might be welcome if more of their expenditures benefited local people. Due to natural fluctuations in ecosystem functions etc. bio-diversity constraints can also suddenly become more limiting. But investments can also be made in order to increase a site's carrying capacity. Technological or policy innovations may also prompt more efficient use of resources and thereby redefine environmental limits. Given this, our submission is that a key issue in improving the sustainability of production and consumption systems in the tourism sector is the need to understand the role and well-being of host communities as well as their support and goodwill in the tourism product, along with attention to the more conventional production and consumption subsystems.

Coastal Tourism in Goa

Coastal tourism is a particularly appropriate domain of knowledge and action to understand sustainable PCS as this is the activity which is at the interface of humans, land and water. Coastal tourism offers a rich arena to understand the play of human agency with ecosystems, through an assessment of the nature of consumption involved (Noronha 2004). Coastal tourism, at least in the tropics, has its triad of central attributes of sun, sea and sand. Given that access to sea and sand is limited in the rainfall months, a certain seasonality accompanies this form of tourism that shapes the choices made in the tourism industry. Society organizes its production and consumption choices depending on the activities possible during different period. For example, infrastructure is scaled up to meet peak demand, which implies that while there is an additional stress on coastal resources during these times due to heavy tourist pressure on limited space and over a short time, this infrastructure mostly remains underutilized in the off-season.

Tourism Development

This case study draws on our research in coastal tourism in Goa, India (Noronha and Nairy 2003; TERI 2000). The research focused on the nature of the tourism driver, the kinds of incentive structures it creates, the impacts on human behaviour and the resulting environmental outcomes that follow. We do not however hold the view that tourism should be treated as an external variable which leads to a series of environmental changes, while the host region is passive and receiving. Our research showed that the internal social, economic and political conditions interact with tourists in flows to result in ecosystem transformations. This implies that we adopt a perspective that sees tourism as being able to be localised in the region, rather than solely as an agent of development or dependency.

The study was located in Goa, a small state on the northwestern coast of India with possibly the longest colonial history in Asia (1510–1961). Goa was put on the international tourist map in the seventies by the 'flower children' or 'hippies' (Noronha et al. 2002). Since then the tourism image of Goa, while catering to a wider range of tourists, remains one of emphasizing sensual pleasures. Tourism is an important industry in Goa contributing directly and indirectly around 20% of the state's net domestic product.[2] The dominant reason for tourist attractions in Goa is leisure, social and cultural

[2]These are estimates as no study has been done to correctly assess this contribution.

interest, and more specifically its beaches. Tourism in Goa has been mostly accommodated on the coastline – over 90% of domestic and over 99% of international tourists go to the coastal areas – and can be termed as coastal tourism although we now see an attempt at regional diversification.

Tourism was adopted as a key sector of Goa's development in the mid seventies, not only for the well established reasons of increasing income, foreign exchange, rural opportunities and employment but also because tourism promised to provide employment of the non-manual kind in a state with rapidly increasing educated unemployment and hardly any industrial growth (Sawkar et al. 1998). The State government embraced this activity but without a clear focus on how it should be developed. Today, Goa is a host to all types of tourists both domestic and international, and within international, backpack travellers, charter tourists and independent tourists. Domestic tourists comprise 80% of all tourists. These come in search of the culture that is "different" from the rest of India. There is a degree of mysticism about the "Goan image". A sense of freedom and "unconventional" dress style are part and parcel of this image. The second is the international tourists who visit Goa for the natural environment of sun, sea and beaches. Interestingly each of these types of tourists has concentrated in different pockets in Goa and has been partly isolated from each other. These different types of tourist have different consumption patterns and cause the destination to respond accordingly in terms of infrastructure and services that it provides.

Since the late seventies, the growth of tourism has been rapid. In the eighties, state government policy supported charter tourism. Despite the international image of Goa's tourism, the share of domestic tourists has always been much higher than that of international tourists (Fig. 13.1). The share of international tourists has

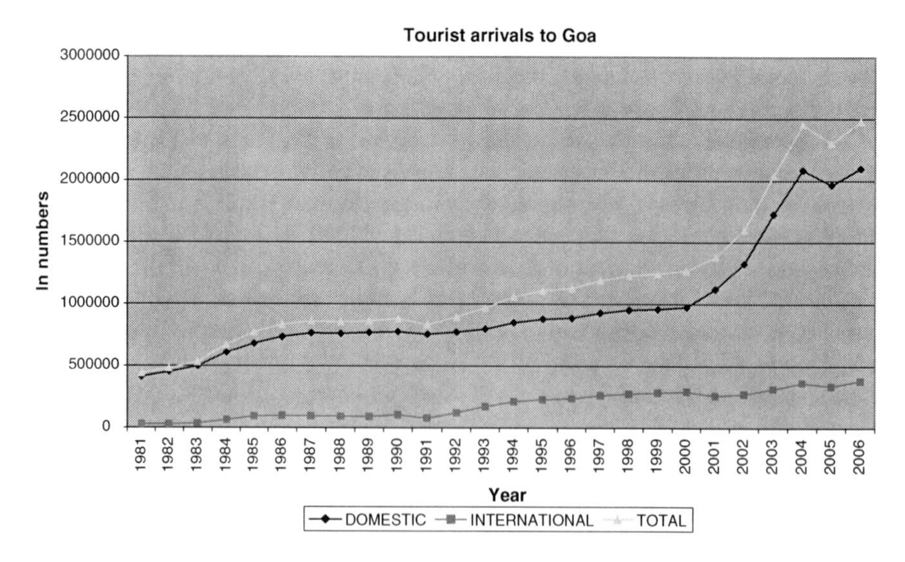

Fig. 13.1 Tourist arrivals in Goa, 1981–2006 (Department of Tourism, State Government of Goa, Annual Reports, several years)

fluctuated over time (Fig. 13.1). In the period 1981–1983 the share was less than 10%, touched 11–12% in 1985–1986, declined in the following three years and showed a continuous rise after 1991, reaching 22% in the later 1990s. The rise in the 1990s was due to the fall in the value of the Indian rupee devalued as part of the structural reform package and diversion to Goa due to terrorism in the competing destinations in Sri Lanka and Kashmir.

Various factors have contributed to the high fraction and continuing growth in domestic tourism in Goa (Fig. 13.1). These include, for example, an expanding Indian middle class with disposable outcomes and cars; advertising by central and state governments as well as the tourism industry; cheap accommodation and resorts; employee benefits for travel; and, time sharing of holiday accommodation (Sawkar et al. 1998). Inexpensive mass transport like the coastal Konkan Railway and South-Central Railway has considerably improved the accessibility of Goa to the rest of India and spawned a form of weekend tourism.

In 1991 the population of Goa was 1.2 million and the annual tourist inflow was 0.8 million, a ratio of 71:100. By 1998, this ratio had risen to 83 (Noronha 2004) and by 2001 to 1.02. Tourist arrivals thus exceed the domestic population and are concentrated in the coastal strip.

Study Focus and Design

Most of our research has been focussed on northwestern Goa, in particular the district of Bardez (Noronha and Nairy 2003). This has six administrative units or *talukas*, and a population of 0.67 million which represents 57% of Goa's population and has a major share of the tourists. Within north Goa, our collaborative research focused on the Baga watershed, and a small micro watershed of the Nerul river, which links our main watershed to the River Mandovi, the main river in the region. Our coastal ecosystem focus was on land, aquifers, and water. The overall approach of the research was cross disciplinary (Noronha and Nairy 2003). It involved different disciplines – biologists, economists, hydrogeologists, sociologists, botanists, coastal oceanographers, and environmental scientists. Different aspects of the problem were studied using customary disciplinary theories and methods. Results were then shared, viewed and reviewed in the context of results from other disciplines, and regular exchanges were maintained between the social and natural sciences.

Ecosystems and Tourists

In earlier research, we noted that it is not tourist inflow per se, but the quantity and type of resources that are used to service this inflow that bring about changes in coastal ecosystems (Noronha et al. 2002; TERI 2000). "Before the entry of tourism, coastal resources served both as production and consumption goods. That is, these resources were used in the production process to earn an income, but they were also used in final consumption, in the sense of the aesthetics and pleasure that these provided the locals. With tourism, it is evident that the villagers have increasingly come to see these resources as means of providing goods and services to tourists"(Noronha et al. 2002).

The effects of tourism upon ecosystems can be better understood if tourism is described in two ways: (i) with reference to the time period in which a destination has had tourist activity; and (ii) the type of participating tourist, which gives a destination its distinctive character as a low-, middle-, or high budget destination (Noronha et al. 2002)

It is not tourism related population movements per se that leads to patterns of unsustainable ecosystem pressures, but the changes in the relationship between people and ecosystems (Noronha et al. 2002). Factors that affect this relationship include host community aspirations and institutional arrangements, such as tenurial and land laws, as well as lack of awareness of environmental services provided by ecosystems. Patterns of out-migration and different forms of tourism also yield different ways of valuing an ecosystem. Al of this has implications for ecosystem use. For instance, overall there may be no *land-cover* change but only *land-use* change (see section on Conversions of Coastal Lands) (Noronha et al. 2002).

One key mechanism to monitor sustainable production and consumption systems is through developing indictors to monitor development activity. In this case, indicators were developed using a DPSR framework. Drivers refer to societal drivers that influence tourist inflows and population movements; pressures are the stresses created by the driver; State are actual conditions of the ecosystems; Impacts are the effects on people, resources and ecosystems; and responses are actions taken by stakeholders (Sonak et al. 2002, p. 368)

Some of the indicators of drivers and pressures in North Goa are summarized in Table 13.1.

Three key impacts of tourism on coastal ecosystems in Goa are conversions of multifunctional land such as the Khazans (see more details below) to built-up area for tourism infrastructure, impacts on ground water and degradation of beaches, sand dunes and their vegetation.

Coastal Land Conversions

Agricultural fields are experiencing a number of changes. This is particularly the case of the *khazan lands*, multifunctional agro-ecosystems in estuarine flood plains that are very productive due to silt accumulation. Using gates and bunds to control water levels and salinity, these fertile systems yield rice, agricultural crops, fish and salt (Noronha et al. 2002). Economic forces are driving change here. On the one hand, expectations of higher returns from the sale of land to

Table 13.1 Driver – pressure – state – impact – response indicators for a tourism destination (Sonak et al. 2002, pp 381–382)

Issues	Indicator type	Results from the study area for 2000[3]	Policy relevance
Social			
Social			
Population movements	Driver	Low migration (net migration rte less than 1.5% per annum)	Employment available
(In-migration, out-migration, tourism-related seasonal in-migration)			Unemployment and environmental stress Availability of jobs
Level of urbanization	Driver	High (more than 40%)	Population pressure and stress ecosystems
Tourist arrival	Driver	High	Stress on ecosystems
Tourist accommodation	Driver	High	Proxy indicators of stress on ecosystems Land conversions
Tourist satisfaction	State	Medium	Response of tourists to involvement and attention from industry
Industry status	State and impact	Medium	Indicates industry satisfaction and (indirectly) future of activity
Local host (community satisfaction)	Response	Good (high)	Indicates community involvement in activity
Economic			
Occupational shifts (movement across economic sectors)	Driver	Primary sector to tertiary sector	Degree of urbanization and modernization
Household earnings	Impact	Rs. 5,291 per day	Command over goods and services
State revenue earned	Impact	Rs. 1,580,000 per day	Importance of the activity and future policy support
Foreign exchange	Impact	2.08 billion rupees	Importance of the activity and future policy support
Human resource training for quality tourism	Response	Low	Awareness of tourism's interactions with other domains
Use of local resources	State	Low	Improved multiplier effects

(continued)

[3] These results are documented here in order to relate the findings to the study area.

Table 13.1 (continued)

Issues	Indicator type	Results from the study area for 2000[3]	Policy relevance
Ecological			
Land-use change (changes in distribution of land under various activities)	Impact	From productive resource base to built-up area	Highlights productive, protective, and speculative use of land Reduced diversity of use for land
Daily withdrawal of groundwater	Pressure	309 L per room per day	Availability of water
Consumption of water	Pressure	617 L per room per day	Availability and wastage of water
Solid waste generation	Pressure	0.92 kg per room per day	Consumption of materials Need for disposal sites
Waste water generation	Pressure	487 L per room per day	Implications for ground and surface water if sewage is untreated
Corporate green practices	Response	Medium	Sustainable resource use and less resource degradation
Governance			
Government initiatives	Response	Satisfactory	Direct links with policy
Certification schemes	Response	Low	Environmental awareness and practices followed
Consumer awareness	Impact	Medium	Good environmental practices
Judicial intervention	Response	Satisfactory	Conflict resolving mechanism

builders and/or from hiring out houses to tourists rather than from actively engaging in agriculture and fishing create incentives for moving out of these occupations. On the other hand, many people have already moved out of agriculture, and are in the tertiary sector, or non-resident, and so have a lower personal dependency on the ecosystem. This coupled with sometimes an insufficient knowledge of ecosystem values and lower, personal (as compared to community) valuation of ecosystems, result in trading in multi-functional lands such as Khazan lands for supporting tourist infrastructure and services such as boating. Khazan lands been converted from having a use that involved the production of paddy, salt, and even coconuts on the dykes, to increasingly having a visual use or to serve as a tourist infrastructure site. These changes have not, however, been a result only of tourism, but also of changes in local political, social and legal institutions over time, such as capital inflows in the form of remittance income, new tenurial laws and changes in the common property systems (Noronha et al. 2002). These are processes, which began in the last part of the nineteenth century and then moved onto a qualitative different and quicker plane post-1961 with the advent of tourism (Siqueira 1999). Zoning rules and poor enforcement, changes in legal rules that had implication for common management regimes, local politics and power configurations further facilitated this transformation (Noronha and Nairy 2003; Noronha et al. 2002).

Coastal Aquifers

Coastal aquifers in the Baga-Nerul watersheds in northern Goa's coast are at risk. Increased demand from population growth (permanent and transient), lack of policy to regulate access and use, technology of extractive equipment, economic activity, insufficient attention to the attributes of the aquifer system and development of a water market has resulted in a medium to high vulnerability of coastal aquifers as assessed using the DRASTIC index and based on evidence of salt water intrusion (Lobo-Ferreira et al. 2003). The chemical and biological analysis of the groundwater samples from open wells in the study area have indicated that the groundwater is heavily contaminated with bacteria and to some extent nitrates. The main source of contamination is presumed to be the sewage disposal sites, the septic tank and soak pit disposals of hotels, restaurants and guest houses. Reports say that sewage was hardly being treated in 99% of low-budget, 100% of middle-budget, 89% of high-budget, and 33% of luxury hotels, sewage was being disposed of in soak-pits or tanks. Only 11% of high-budget and 67% of luxury hotels were able to treat their sewage, as they had treatment plants (Noronha 2005). Models of water resource needs and local aquifers have now been developed to assist with planning for tourism-related infrastructure and to locate wells to secure supply for existing hotels (Lobo-Ferreira et al. 2003). These take into account risks of saline intrusion which rise as proximity to shore decreases.

Degradation of Beaches, Dunes and Vegetation

Tourism seems to have made the area greener but has reduced its previous ecological diversity. While mangroves have marginally increased in terms of real extent, area under sand, coconuts and cashew plantations, and dense semi-natural woodlands have decreased in a number of villages. While the Normalized Difference Vegetation Index (NDVI) – an estimate of vegetation density obtained using remote sensing data – has increased, the species diversity has reduced in the study area with loss of the original vegetation (Feoli and Giacomich 2003); (Jagtap et al. 2003). Population, user behaviour, poor rule enforcement, and lack of technology for waste management have polluted the beaches and reduced dune vegetation as these are cleared to accommodate tourism services.

Host Communities and Tourists

Apart from the environmental issues, a number of social problems have also accompanied tourism. These include increase in drugs, HIV/AIDs and paedophiles. Social problems and conflicts over values have inspired, and been the foundations for, anti-tourism activism, much of which is targeted at international tourists and unplanned growth (Noronha 1999). The more prominent groups include Jagrut Goenkaranchi Fauz (JGF), Citizens concerned about Tourism (CCAT), Saligao Nagrik Samiti and the Goa Foundation. The Goa Foundation, for instance, has maintained campaigns against several large hotels that ignored building regulations and blocked some other proposals by local politicians (Noronha 1999; The Goa Foundation 1989). Local newspapers and NGO submissions to government help to emphasise the public critiques of tourism.

Sustainable tourism is dependent not only on healthy ecosystems but also the support and good will of local host communities. Good relations with local hosts create both comfort and memorable experiences for tourists. Socio-cultural interactions with tourists, some argue, offers opportunities for residents to interact with other cultures or absorb new ideas into the community. Some also view that "the value that tourists attach to indigenous traditions and practices can actually encourage local communities to maintain their unique traditions, customs and identity" (Robertson et al. 2008)

Poor relations with host communities occur if tourists lock themselves away within resorts and do not experience the 'place'. "Tourism, especially when it attempts to cater to the needs of a global tourist attempts to emulate in the host destination, the conditions that a tourist is looking for. This may not necessarily be what the tourist is looking for, since the tourist may in fact be looking for a more 'local, variegated' experience. But it is often assumed, by host populations and governments, that the greater the reconstruction of the conditions that a tourist is accustomed to, the higher will be its attractiveness as a tourist destination" (Noronha et al. 2002).

Current practices in tourism development are partly to blame for severing this valuable relationship in Goa. A tourism development path is being followed that seeks to reproduce for the international tourist his/her home conditions and for the domestic tourist the conditions that emulate the west, instead of adopting a path that will provide the tourist a more 'diverse experience'.

With modern mass tourism and the goods and services that are being developed to cater to this type of tourist, tourist villages are orienting themselves toward production of more homogenous outputs. Arts and handicrafts, for instance, are being manufactured for bulk sale and consequently lose some of their quality (Sawkar et al. 1998).

Shortages of electricity and water may occur: the residents of coastal Goa's Candolim and Calangute villages, among the first to be involved with tourism since the 1960s, have complained that the water pipeline originally meant for the villagers has been taken over by the hotels (Noronha 1999).

Another social issue is the seasonality of income and employment, affecting smaller hotels and unskilled workers the most (Sawkar et al. 1998). Changes in livelihoods have also been substantial, particularly in villages along the coast. Fisheries and agriculture have been replaced by tourism-oriented occupations.

Land prices have risen sharply and thus land has been subject to speculation with officials alleged to be involved in corruption (Noronha 1999). Some of the most over-built villages have become very ugly and inconvenient places for local communities to live in.

Finally there are also significant contests between different resource users even within the tourism sector, for example, the ongoing conflicts between larger and permanent hotels and resorts and smaller, more temporary, beach shack owners (Noronha 2004).

Discussion

The socio-ecological basis of coastal tourism in Goa has not been adequately recognized. Several existing and looming environmental and social problems are not being effectively addressed by policy despite some significant scientific understanding of trends, status and causes. Research-based knowledge has not had much influence except where it has been taken up by advocacy groups. Tourists themselves, the key consumers, have largely been missing from analysis of opportunities and responses, in particular the very large fraction of domestic tourists. Future promotion of tourism should consider much more closely impacts on, needs of, and changes in, the supporting social and ecosystems.

The knowledge most needed for PCS in tourism in Goa is a better understanding of the ways in which pressures are created, the agents and processes involved in the creation of those pressures, and ways in which this knowledge can be converted into action. Not all of this knowledge is actionable, as some of it is much too endogenous and systemic, such as for example, pressures for conversion of Khazan lands under the demand for land for tourism.

On the sustainable consumption side, there is a need to work with the tourists' expectations and perceptions which can be modified by active communication and by the destination's marketing policy. Not much attention has been given to the opportunities for improving sustainability through greater attention to interests and motivations of different kinds of tourists in Goa. Some of the dominant assumptions among producers may be misplaced.

To ensure sustainable production, the highest priority actions to narrow gaps between knowledge and action as they emerged from our case study are in the policy front. There is a strong need for better enforcement of existing environmental legislation and plans. Regional development plans of the Government of Goa look fine in print "the location of new beach resorts should be considered not only from point of view of land availability but also from the consideration of beach resource ecology" (Sawkar et al. 1998) but capacities of agencies to act or take responsibility are often lacking and there are inadequate benchmarks against which to judge good and bad practices.

Local bodies face particular challenges as they are expect to implement rules which conflict with needs and aspirations with the result that coastal tourism growth has been largely uncontrolled (Sawkar et al. 1998). Thus, in anticipation of central government implementation of coastal zone regulations (CRZ) in 1991 that would restrict construction in the coastal areas, the state government encouraged and sped up approvals for coastal infrastructure projects (Noronha 2004). Clear violations in later years continued to be tolerated. There is a need for review of land tenure legislation and the introduction and implementation of zoning policies.

Some of the most serious impacts from tourism are on water resources. Better policies are needed to control ground water use and quality and integrate these with management of surface waters. Use has historically been unregulated (Noronha 2004). Research reviewed earlier suggests looming quantity and quality problems. Planning that leans on notions of carrying capacities should take into account both social and ecological dimensions and be wary of the inherent dynamics.

Greater involvement of the local host communities affected by tourism developments in the public policy process is perhaps the most critical need for governance. A history of little public involvement in decision-making related to tourism planning and policy helps explain some of the alienation felt among host communities about tourism related changes (Sawkar et al. 1998). Another related issue is a set of private actors with high stakes in tourism industry including hotels, casinos, pubs and discos that are notoriously powerful in the political economy of Goa (Robertson et al. 2008). Such special interest groups will pose challenges to attempts at more open and broader public consultation that seems to be needed for dealing with the complex and unstructured issues associated with planning sustainable coastal tourism (Noronha 2004).

We suggest that sustainable PCS in tourism requires market conditions for both consumers and producers of the tourist product to change. We believe that a more participatory form of tourism could provide a framework for promoting sustainability of tourism PCS. Different stakeholder groups would then plan for tourism that addressed the needs of all groups; monitor impacts and manage change (i.e. drivers) to avoid arriving at carrying capacity limits; and maintain or improve tourist–host community relations. Wider stakeholder participation in planning and developing tourism destinations should be able to address social and environmental ills and risks of tourism while also garnering and distributing its benefits and helping resolve conflicts.

References

Feoli E, Giacomich P (2003) Land cover patterns. In: Noronha L, Lourenco N, Lobo-Ferreira J, Lleopart A, Feoli E, Sawkar K, Chachadi A (eds) Coastal tourism, environment, and sustainable local development. TERI, New Delhi

Jagtap T, Desai K, Rodrigues R (2003) Coastal vegetation patterns in a tourist region. In: Noronha L, Lourenco N, Lobo-Ferreira J, Lleopart A, Feoli E, Sawkar K, Chachadi A (eds) Coastal tourism, environment, and sustainable local development. TERI, New Delhi

Lobo-Ferreira JP, Chachadi AG, Oliviera MM, Nagel K, Raikar PS (2003) Groundwater vulnerability assessment of the Goa case study area. In: Noronha L, Lourenco N, Lobo-Ferreira J, Lleopart A, Feoli E, Sawkar K, Chachadi A (eds) Coastal tourism, environment, and sustainable local development. TERI, New Delhi

Noronha F (1999) Ten years later, Goa still uneasy over the impact of tourism. Int J Contemp Hospital Manag 11(2/3):100–106

Noronha F (2005) North Goa feels the tourist pinch. People and eco tourism. Third World Network Features, Malaysia

Noronha L (2004) Coastal management policy: observations from an Indian case. Ocean Coastal Manag 47:63–77

Noronha L, Nairy KS (2003) Changing uses, ecosystem valuation, and perceptions: the case of khazans in Goa. In: Noronha L, Lourenco N, Lobo-Ferreira J, Lleopart A, Feoli E, Sawkar K, Chachadi A (eds) Coastal Tourism, Environmental, and Sustainable Local Development. TERI, New Delhi

Noronha L, Siqueira A, Sreekesh S, Qureshy L, Kazi S (2002) Goa: tourism, migrations, and ecosystem transformations. Ambio 31(4):295–302

Sawkar K, Noronha L, Mascarenhas A, Chauhan O, Saeed S (1998) Tourism and the environment: case studies on Goa, India and the Maldives. World Bank, Washington, D.C

Siqueira (1999) Global Consumption and Local Processes. Report submitted to Tata Energy Research Institute, Goa. TERI, Goa

Sonak S et al. (2002). "Driving force, pressure, and impact indicators for a tourism destination", in "Coastal tourism, environment, and sustainable local development", ed L. Noronha, N. Lourenço, J.P. Lobo-Ferreira, A. Lleopart, E. Feoli, K. Sawkar, A.G. Chachadi, 464 pp, TERI (India)

TERI (2000) Population, consumption and environment inter-relations: a tourist spot scenario. Tata Energy Research Institute Project Report No. 97EM50. TERI, New Delhi

The Goa Foundation (1989) Sand mining, beach resorts and the destructions of sand dune ecosystem in Goa. Goa Foundation – studies on the Goan Eosystem-I, Mapusa, Goa

Chapter 14
Conclusion: Linking Knowledge and Action

Louis Lebel and Sylvia Lorek

Introduction

Transforming production–consumption systems (PCS) is fundamental to the pursuit of sustainability in an interconnected world (Chapter 1). Diverse experiments are underway. These include efforts to provide services rather than sell or transfer ownership of products, green supply chains, jointly design products with consumers, standards and regulations to support fairer trade and ethical marketing, and campaigns for responsible buying and using less (Lebel and Lorek 2008).

The pursuit of sustainability is knowledge intensive. A recurrent challenge is to bring knowledge, or *justifiable belief* (Rorty 1979), and action closer together. By actions we mean "*doing something that has physical or behavioral repercussions*" (van Kerkhoff and Lebel 2006). Experience, learning by doing, and knowledge embedded in current social practices like habits and lifestyles may each be important. The links between knowledge and action, thus, are often much more complex than direct linear flows of information from research to policy then practice. They are more like *arenas* in which people interact, sharing knowledge, exercising power, and taking actions (van Kerkhoff and Lebel 2006). Arenas can be physical or virtual places.

Knowledge–action gaps have several causes. The knowledge needed to transform unsustainable production–consumption systems into sustainable ones may be missing, inaccessible, or available but not acted upon. Gaps may arise from both the knowledge and action sides of a problem as well as combinations.

In this final chapter we review what has been learnt about what it takes to effectively link knowledge and action to improve the sustainability of PCS. Our primary goal

L. Lebel (✉)
Unit for Social and Environmental Research, Chiang Mai University, Chiang Mai, Thailand
e-mail: louis@sea-user.org

S. Lorek
Sustainable Europe Research Institute (SERI), Head of Sustainable Consumption Research, Overath, Germany
e-mail: sylvia.lorek@seri.de

L. Lebel et al. (eds.), *Sustainable Production Consumption Systems: Knowledge, Engagement and Practice*, DOI 10.1007/978-90-481-3090-0_14,

is to identify effective strategies of actors and promising supporting institutions. We draw on the findings of the case study contributions in this book and enrich these, where appropriate, with related published work.

The rest of this chapter is organized into three main parts. The first synthesizes what is known about knowledge–action gaps, their forms, causes and consequences. The second summarizes what actions, strategies and institutional innovations have been tried to close gaps and how they have performed. The third and final part reflects on the wider significance as well as limitations of this analysis of PCS.

Gaps

At a general level the kinds of knowledge *needed for* making PCS more sustainable includes things like ways to: increase the efficiency with which physical resources are used; reduce the amount of waste; increase the amount of utility or benefit; and improve fairness in how burdens, benefits and risks are allocated. The specific knowledge requirements, and the ease with which they can be pursued, vary hugely among PCS, which actors and relationships are under consideration, and circumstances. The forms and causes of gaps identified in the case studies reported in this volume were diverse. Knowledge–action gaps can arise because knowledge is missing, inaccessible or unused and actions are therefore ignorant, poorly informed or irrelevant.

Missing

Knowledge may be missing because there is no research, or research has been carried out but without discovering salient solutions. A lack of research may be a result of ways agendas are set and lead to insufficient investment. Some problems in sustainability are inherently complex and, despite substantial research effort, may remain intractable until the research questions are reframed.

Agendas and Investments

Problems of insufficient investment explain why some knowledge is missing. Some important topics are simply under-researched. Actors with needs may have no influence or authority over research and action agendas. A large fraction of the world's population does not consume enough. The poor can only buy what is cheap, and what is cheap is often very low quality. Making right choices is doubly hard when consumer rights are not protected and there are insufficient resources for awareness or education programs (Chapter 4). Consumer activism in developing countries is difficult and must pay attention to poverty eradication, quality of life and well-being (de Zoysa 2005). There is not much research or assessment activity framed to deal with issues from developing country consumer perspectives. The international programs on sustainable consumption have largely ignored issues of under-consumption and poverty (Chapter 4). This is critical as most developing countries lack bargaining power in international negotiations on trade, financing and the environment. For these countries issues of food security are tightly linked to maintaining ecosystem services from diverse agricultural systems. Research on sustainable livelihoods comes closer to meeting these knowledge needs, but would benefit from greater interaction with methods and findings typical of PCS research carried out in developed economies, especially with respect to commodity chains for local consumption.

Research in reducing wasteful consumption in the wealthier parts of the world also suffers from strait-jackets. Work in the corporate sector, in particular, has followed very traditional lines. Very little is known about the parts of business, for example, which support shifts to lower consumption but higher prosperity (Chapter 2). Tourism research has remained destination-oriented and largely left questions of consumption volume aside (Chapters 11 and 12). Opportunities for leverage further upstream, like at travel agencies, where tourists make decisions about destinations, and consider alternative ways to spend leisure time have been much less explored. High carbon taxes for air-travel may provide an incentive for changes around this relationship. Low and lower consumption research agendas need to be reframed to give more attention to firms in the private sector and how they interact with employees and customers.

Research priorities and whose knowledge counts can change with circumstances. The energy crisis in the 1970s led to some support from the German Government for research in renewable systems, but the main initial push was towards nuclear power (Chapter 3). Under public pressure and concerns about nuclear greater support for research and development on renewable energies followed in the 1980s. An early example beginning in the energy crisis era was the research and test facility under the GROWIAN project focused on very large scale, high technology, wind turbines (Chapter 3). The project involved government, wind energy experts and big business coalition. But it largely failed. Citizen-led initiatives focusing on smaller-scale, less radically innovative, but practical technologies were initially more successful. In the 1990s regulatory changes, such a feed-in tariff system saw market-oriented larger businesses rather than cooperative model in wind energy predominate. This case study illustrates clearly that social values and political decisions shape technology and related infrastructure development which in turn influence consumption (Chapter 3).

Inherent Complexity

The inherent complexity of sustainability challenges can make it difficult to carry-out salient research. Thus, the social and ecological carrying capacity of tourist destinations is often not well understood (Chapter 13) and consequently easy to ignore in decision-making (Chapter 11). One reason is the ecosystem changes that arise in tourism destinations are a product of interaction between tourist flows and various social, economic and political factors (Chapter 13).

Likewise the full set of environmental and social impacts and benefits along complex and long commodity chain are often not well understood. In the case of shrimp (Chapter 7) the focus has been limited to environmental impacts of outgrowth ponds in developing countries or on health risks to wealthy consumers of chemical residues; social aspects have been largely neglected with little known about working conditions or resource access issues for local communities despite some early reports suggesting they were potentially very important (Bailey 1988).

The chain or network structure of PCS itself can add complexity to the problem of finding appropriate knowledge. Thus, a policy aimed to stimulate innovation or adoption in a particular part of a PCS may not have much use for processes one step removed from those targeted, but still have great impacts on them, resulting in compatibility problems, especially where the corresponding "innovation" at the next or prior step simply isn't available. This is an additional "landscape" (*sensu* Geels 2005) constraint to innovation in PCS where there are tightly coupled chains or networks. For example, the Thai aquaculture industry originally based on culture of native Black Tiger Prawns struggled despite substantial research and development investments to fully domesticate the species, continuing to depend on wild-caught gravid females to supply hatcheries (Chapter 7). This meant much less control over genetic stock and made disease management much more difficult compared to the exotic Pacific White Shrimp which, once permission was granted to introduce, took over as the preferred culture species. The switch to white shrimp was accompanied by parallel changes in feed composition, number and organization of hatchery firms, and introduction of new standards and certification schemes. While most innovations in PCS are probably incremental and local in influence others may arise that could transform a large number of components relatively quickly because of inter-dependencies.

The expansion of agrofuels triggered a complex set of interactions between what had been relatively independent energy, food and agrofuel PCS (Chapter 6). Energy prices now have major impacts on agricultural markets. Understanding of these relationships remains very preliminary and given the larger number of factors involved future research will likely have to continue to deal with dynamic uncertainties as part of its analyses.

Closing the gap between knowledge and action when knowledge is missing implies: investing in and doing more research, and re-framing problems into something solvable, for instance, by incorporating uncertainties and allowing for non-linear relationships.

Inaccessible

Knowledge may be inaccessible because of problems of communication or understanding. Access difficulties may also arise from insufficient capacities to acquire knowledge.

Communication

Problems of poor communication and insufficient capacity largely explain why some knowledge exists but remains inaccessible. Transparency, for example, is a key issue in the pork chain in Europe (Chapter 9). Firms producing organic products are keen to promote codes and standards and education which reinforces their nutritional and lifestyle messages, but other mainstream firms have less interest to do so, especially about non-product qualities. The presence of too much information on what is good nutrition and so on, leads to both insecurity and confusion, making quality communication even more important (Chapter 9).

Observations at tourism sites in Thailand illustrate the challenges in getting even relatively simple messages about preferred or expected behavior across (Chapter 12). Signs need to be in multiple languages, they need to be visible and persuasive. Many hotels, find simple signs appealing to customers not to waste water effective in reducing towel use and thus cutting their expenses. In some cases signs need to be supported by more individual interventions as infringements occur, as in the case of whistle-blowing guards who keep watch for culturally inappropriate behavior in temples (Chapter 12).

Many consumer activists hope to close knowledge–action gaps with more information. Educating consumers is a common response to the perceived problems of consumerism and as a way of taking cautionary understandings into action. Most consumer education campaigns start from a rather narrow base of assumptions about lifestyles and values; this is true regardless of whether the origins are with state agencies, for profit or non-governmental organizations (Seyfang 2004; Knight 2004). Meanings and content of attitudes and behaviors change over time and with lifestyle dynamics (Chapter 3). Green consumers are also constructing a particular image about themselves through the commodities they buy (Connolly and Prothero 2003) and this can be used to advantage by firms with greener products. Even so there is often a profound gap between values and actions (Burgess et al. 1998). As a consequence just providing more information may not lead to behavioral change (Hobson 2003).

For electricity production the knowledge problems are largely not of a fundamental nature about impacts, but rather on how to accurately communicate information about the environmental relevance of the energy mix to consumers on the one hand, and how to shift production more towards renewable energies through regulations and incentives by governments on the other (Chapter 5). Reliable information on energy mix for consumers is hard to get. Electricity suppliers should provide the information. But even higher leverage would come from supply-side energy mix improvements that require government intervention and not just reliance on consumer purchasing behaviour (Chapter 5).

Appropriate technologies or understanding may exist in minds, firm or places, but still far from those in need. Communication failures may arise because information is in the wrong form to be accessed and used, its saliency unrecognized. Knowledge itself may conflict with other core beliefs or habits and thus be hard to introduce in practice. There are both accidental as well as intentional barriers to communication and understanding of research-based knowledge.

Capacities

Producers, consumers and other actors involved or affected by a PCS differ with respect to capacities for creating, acquiring or keeping knowledge and for taking different kinds of actions.

Producers usually have more knowledge about their production process, site and products than others not so intimately involved in those activities, but that knowledge may not extend to environmental performance and sustainability concerns. They also have capacities, responsibilities and freedom to act. Producer responsibility has two elements. The first is to take the responsibility into action in their production process and for the products. Here they have undisputed power to act. The second is to report about it and so create knowledge and transparency for other actors along the product chain; producers can also be knowledge transmitters. (cf. Chapter 9). Producers form associations which can have important functions in sharing knowledge about new ideas, like sustainability or environmental standards (Chapters 7 and 8).

Consumers often have little knowledge about sustainability from their own experience. Whatever they know is learned or just heard. Therefore it is crucial to provide them with trustworthy and salient information (Bilharz et al. 2008). Consumers may preferentially buy what others thought to be more sustainably produced. Consumer organizations, for example, independently evaluate products and organize campaigns to raise consumer awareness of issues or boycotts against particular products or firms. The avenues for consumer activism, however, may be restricted in non-democratic, developing, societies (Chapter 4).

Workers need to be distinguished from the firms that employ them. They have specific and detailed practical and tacit knowledge about production processes that can be used to improve routines. Their capacities to use and share knowledge are an important part of PCS. Moreover, they may bargain for better conditions but without agreements with management first their possibility to act on knowledge are often highly constrained. Furthermore, many workers in wealthier countries are interested in trading reductions in income for leisure time (Chapter 2) and so further links between production and consumption might be made. A challenge is getting firms to support such actions (Schor 2005).

Affected communities living around a production site can report how production influences their lives. Communities living on Pulau Langkawi, for instance, were displaced to make way for tourism development projects, lost access to beaches cordoned-off by hotels, and have seen their needs largely ignored by government infrastructure projects (Chapter 11). Some communities in tourist destinations lobby firms to reduce adverse impacts or improve beneficial ones (Chapter 12). Research-based knowledge in Goa has not had that much influence unless it was taken by advocacy groups (Chapter 13).

Traders and retailers like supermarkets, travel agencies or sea-food processing firms typically know things about a product's characteristics and where it came from. They may provide information to consumers via labels or relationships of trust (Chapter 8). They can demand certain sustainability criteria be met and optimize distributional and exchange logistics in their self-interest (Chapter 10).

Regulators help shape systems of governance at different levels of social organization. They include state and international agencies concerned with regulation and monitoring of production and trade (Chapters 5 and 6). They often have working knowledge about policy integration opportunities and possible side-effects. They are also cognizant of implementation challenges, for example, in enforcing regulation and monitoring compliance. Key actions include revising laws and regulations, giving penalties and rewards, and gaining commitment to principles. But at local levels capacities of governments to acquire information and use it for planning maybe quite limited, as for example, in tourism site development in Thailand (Chapter 12).

Researchers use systematic methods to uncover new forms of, and validate existing, knowledge. They develop better technical procedures – to use resources as well as govern social behaviour. Research creates an enormous amount of knowledge. Compared to that, their ability to act is rather limited. Scientists have to find other actors to work with them, to interest them and to get them to adopt solutions. This can be as consultants, as members of public goods organizations, or through advocacy.

The capacities to produce, use and access knowledge of different actors varies widely within and among PCS. Identifying the key actors with knowledge about sustainability and mapping with whom they interact can help with diagnosis of knowledge–action gaps in PCS.

Closing the gap when knowledge is inaccessible implies creating new connections between actors that allow shared understanding, and building capacities to acquire knowledge and act on it.

Unused

Knowledge may exist but remain unused because the incentives for its adoption and change in practices are inadequate. Vested interests and power relations may prevent actionable knowledge from being used. Key knowledge claims may be legitimately contested, but nevertheless, through gridlocks prevent informed decision-making and actions.

Incentives

In influencing consumption practices sustainability values must compete with many other criteria that are usually ranked as more important, like price and convenience (Chapter 5). In the German electricity PCS there is a positive attitude towards renewable energy sources among consumers including a stated willingness to pay more. But when it comes to action, difficulties in figuring out actual energy mixes in different options, reduce the ease of action. Highly motivated, green consumers perceive direct personal benefits of their actions but the majority do not see sufficient incentive for making the switch (Chapter 5). The lack of perceived personal benefit is one of the major causes of the knowledge–action gap.

The production and consumption of agrofuels in Thailand is strongly influenced by prices relative to conventional fuels while expansion is limited by the non-availability of land planted to feedstock (Chapter 6). When fossil fuel prices are cheap agrofuels can only become competitive with the aid of government subsidies. The rationale for supporting agrofuel expansion, whether it is energy security, climate protection or other perceived social benefits, has rarely been articulated clearly in individual policies. Targets are set but their relevance to broader goals and their justifications remain mysterious. Cautionary research about the environmental and social impacts of agrofuel expansion, the benefits and costs of full life-cycle based on alternative crops, and the overall modest contribution agrofuels could make to renewable energy supplies are ignored (Chapter 6). Policy-makers have not drawn much on international or domestic research apart from technical research on extracting and processing fuels and highly selective economic benefit studies.

One of the simplest and toughest initial obstacles to many initiatives related to sustainable consumption is the widely held belief or presumption that any reduction in consumption equates to sacrifice or diminished happiness (Chapter 2). Contrary research findings have had little impact on this view (Jackson 2005). Maniates (Chapter 2) argues there are still many opportunities for researchers to reflectively engage in activist agendas, for example, that demonstrate increased happiness with less material consumption.

The attractiveness of sustainability movements, in part derives from their differentiating effect or micro-radicalism. Once fair-trade coffee or organic pork becomes a mainstream product it can be much harder for campaigns to market it to as being specially made for you. It is also gets more difficult to address genuine concerns of consumers about fairness and sustainability when food safety and value-to-health issues dominate concerns of global agro-food networks (Raynolds 2004). Consumers and producers may find it a challenge to be both pro-sustainability and mainstream.

Interests

Powerful, vested, interests may make knowledge for sustainability unusable. Pressures for coastal development for tourism are crucial factors in the unsustainability of these PCS. The often weak political influence of affected local communities means that knowledge about sustainability is not sought (biodiversity, ground water) and what exists is ignored and circumvented. Some knowledge is simply not actionable because of other more powerful factors. In the case of Goa, strong demand for conversion of Khazan lands for tourist site development completely over-rode other concerns and knowledge (Chapter 13). Knowledge that is already institutionalized in regulations may still be ignored as when there is lack of enforcement of the rules on paper (Chapters 12 and 13).

The impact of interests may also be more subtle. Life cycle and economic analyses of agrofuel options has had very little influence on policy-making in Thailand (Chapter 6). Instead decision-making appears to have been primarily driven by short-term concerns with rising oil prices and energy security as well as a means to handle sharp fluctuations in prices of key commodities that are also biomass fuel stocks. The important interests in this case include both agribusiness and energy firms and associated government departments, frequently headed by people with previous private sector experience. Government targets, because they have not been closely related to actual demand or market conditions, have often been way off the mark (Chapter 6). There is a need to both expand scientific and evidence-based assessment of options and to put this more prominently into public domain. Agrofuel projects also need greater independent public scrutiny and assessment before they are approved (Chapter 6).

Food producers in developing countries, especially smaller farms, often have substantial constraints on getting knowledge that reflects their interests into trade negotiations, standards or certification schemes (Chapters 7 and 10). For tourism the relationships are even more complex as many of the tourism service providers may be owned or invested in by firms outside the developing country destination. Local community interests in tourist destinations, however, as in the case of food production, are also difficult to integrate into overall PCS beyond pointing to benefits of job opportunities (Chapter 12).

Contested

For many issues arising in efforts to transform unsustainable PCS, knowledge claims are contested. Credible, salient and legitimate knowledge may be resisted by vested interests, but in other cases it is simply that there are uncertainties in understanding, different approaches and metrics, and alternative sources of knowledge which when brought together don't fit neatly with each other. Uncertainties about benefits, burdens and risks may support caution and procrastination, and inaction.

For example, different certification schemes for coffee have different criteria for what constitutes "sustainability" (Chapter 10). In some cases the focus is on minimum standards, like prices for small farmers, whereas in others the goal is more to support a standard in which the market can operate freely. The presence of multiple standards or labels can create difficulties for both producers and consumers.

In the case of shrimp aquaculture an extensive dialogue process to set "best practice" guidelines has led to some convergence on principles (FAO et al. 2006) of what constitutes acceptable rearing and water management practices (Chapter 7). But shrimp farmers and local communities often still feel that too much of the knowledge being used is generic and does not reflect conditions on the ground; local knowledge and experience, they feel, is excluded by these standardization processes (Vandergeest 2007; Lebel et al. 2008).

In the pork PCS many actors have struggled to move from food safety concerns to consideration of animal welfare and environmental sustainability (Chapter 9). In many of these food PCS, available knowledge about sustainability is selectively used because of concerns with costs and competitiveness. This appears to be true regardless of whether the chain is local, regional or global.

In some cases more direct confrontation with current practices was needed to get wider policy attention. Individual supporters of wind energy in Germany went ahead with their pilot projects independently of larger government-funded research and development efforts (Chapter 3). Consumer groups in developing countries are starting to push beyond traditional circumscribed issues of consumer protection to tackle more problematic issues of trade liberalization, corporate social responsibility and poverty (Chapter 4).

Closing the gap when available knowledge is unused implies increasing incentives for adoption and fostering processes that lead to active engagement of, and deliberation and contestation among, diverse stakeholders.

Closures

A range of examples of knowledge–action gaps were identified in each of the case study domains explored in this book (Table 14.1). Gaps, as we saw, are not necessarily a broken link from knowledge-to-action, but can just easily be in the other direction or a more complex set of relationships requiring understanding of power and interests in the various arenas in which different groups of actors meet. In section we look more closely at evidence about the specific activities, strategies and institutions involved in gap closures (Table 14.1). The discussion is organized under three themes: integrated, deliberated and validated action.

Table 14.1 Examples of closure strategies for common knowledge–action gaps discussed in this volume

	Gaps	Closures
Case Study Theme	Missing, inaccessible and unused knowledge	Integrated, deliberated and validated action
Lifestyles	Good information doesn't lead to changes in practices	Reframe restraint as opportunity not sacrifice; give feedback on performance
Energy	Information about socio-environmental relevance or economic implications not easily available to consumers	Require producers to make it transparent and available; regulate for better mixes overall
Food	Traceability system works but information relevant to sustainability lacking	Use system for food safety simultaneously for sustainability information
Tourism	Interests, issues and beliefs of local communities ignored in decision-making	Dialogue to ensure inclusion of all stakeholders to negotiate more appropriate developments and practices

Integrated

There are several pathways and approaches to integration (van Kerkhoff 2005). For PCS these typically involve: boundary management; labels and standards; and, spanning innovations.

Boundary Management

In most of the case studies in this volume there were actors trying to enhance knowledge–action linkages for sustainability. In the longer-term individuals can influence PCS both as consumers and citizens (Chapters 3 and 4). Although there are often many constraints on choice and "sovereignty" in the longer-term their actions can influence the institutional setting in which production takes place.

Several insights about the management of the knowledge–action boundary are apparent in the "take back your time" and "right2vacation" campaigns (Chapter 2). The "Right2Vacation" campaign argues "*that more paid vacation time will lead to lower work stress, reduced binge vacationing, higher levels of local civic participation, deeper connection to (and appreciation of) local and regional environmental assets, and a growing political awareness of the benefits of trading income (and consumption) for leisure*" (Chapter 2). With this line of reasoning, production and consumption side issues are directly linked. The issue has been reframed in a way that is positive, countering conventional views of less consumption as sacrifice. Intriguingly the campaign has attracted attention from firms that benefit from people having more leisure time, suggesting unexpected relationships once an integrated perspective is adopted. Either way, building coalition to support such campaigns is crucial and requires a shift from thinking about individual to collective action.

The sustainability of electricity PCS in Germany does not result from lack of knowledge, but limited willingness of consumers to take on responsibility for their decisions (Chapter 5). Improved communications that illustrate personal benefits may help, but ultimately it will also take government intervention to reduce transaction costs and tackle supply side issues.

For organic agriculture in Germany, government promotion and regulations have played a role in managing boundaries between producers and consumers. The Ministry of Food, Agriculture and Consumer Protection, for instance, has been active with investments and standards (Chapter 8). But it has been the private sector which has taken the primary lead in summarizing and promoting information about sustainable agriculture, healthy nutrition, and tasty food (Chapter 8). As observed in several chapters in this volume some of the contention among actors is in refining what exactly should constitute "sustainability" in a particular PCS; it may be that operationally, this will often need to be negotiated (Kates et al. 2005).

Agrofuel policy in Thailand suffers from administrative fragmentation among different government ministries (Chapter 6). Attempts to solve this boundary problem through a plethora of agrofuel-related committees appear not to have really solved the problem of managing boundaries between different parts of the bureaucracy that need to share knowledge and make decisions.

Boundary management in PCS, is thus in part about translation and mediation (Guston 2001; Cash et al. 2003), but it is also about linking actors already in business relations but with new, sometimes co-produced, shared knowledge.

Labels & Standards

Getting information about the sustainability of production processes to *stick* to agricultural products is a recurrent challenge in food systems.

In the past market exchanges and value-added chains often depended on trust developed between those involved. With wider number of suppliers and buyers and much more distant and complex chains, dominant firms have sought other means of achieving quality assurance (Renard 2003). In doing so tacit knowledge has been largely replaced with codified knowledge in the form of standards and labelling schemes. The shift in how knowledge and action are linked in general has not been totally accepted by those actors pursuing sustainability. Indeed many of the social movements favoring more local systems of production are also trying to reinvigorate validation of the tacit and informal sort based on networks of trust.

Support for organic networks through public awareness campaigns about the merits of organic production are not on their own sufficient. Knowledge about organic agriculture is still not that well integrated with the retail food sector (Chapter 8). Economic, social and political integration strategies are needed to compliment and reinforce information-based initiatives (Chapter 8).

Information transparency with respect to food safety concerns dominate over sustainability ones in pork and shrimp chains (Chapters 7 and 9). Simultaneously promotion would be efficient. But certification and labeling schemes have limitations. In the case of coffee and shrimp, certification largely changes producer, but not consumer, behaviour (Chapters 7 and 10). Only in very specialized niche markets where premiums are paid – like, fair trade coffee or organic shrimp – does the consumer have greater direct influence.

One of the recurrent challenges for certification schemes as integration instruments for PCS is the extent to which they serve niche or mainstream interests (Lebel et al. 2008). On the one hand, schemes that remain highly constrained to a particular niche are not going to bring about the wider transformations needed for sustainability; on the other hand, mainstream schemes must comply with dominant, conventional, expectations and needs of large players and these may be very hard to shift towards sustainability (Chapter 10). Costs are another key factor limiting broader certification schemes (Chapter 9). From the knowledge–action perspective it appears that the key arenas for these two situations will likely evolve and remain parallel rather than converge.

Spanning Innovations

Linking knowledge and action in PCS has some peculiar features. Many of these arrive from the structure of PCS in which there are many diverse actors, but also shared interests and incentives to improve system-wide performance. Performance, for many actors, however, means profits or even throughputs rather than sustainability.

The Thai government has played a crucial role in the expansion and competitiveness of the shrimp aquaculture industry including investments in research at public and quasi-public institutions (Chapter 7). Over time, notions of sustainability have penetrated broader policy initiatives, partly in response, to the concerns and demands of consuming countries. The involvement from the beginning of a very large and vertically integrated multinational, Charoen Pokphan (CP) Group, has made system-wide what would otherwise be piecemeal innovations, forcing other firms and farmers to follow in their wake. Integration, therefore, has a flipside: it can lead to greater control by single, powerful firms (Goss et al. 2000; Green and Foster 2005).

The vast majority of research effort for sustainability is directed at individual nodes of PCS. The amount of leverage available is correspondingly modest. Most innovations do not lead to major transformations; to do so they must engage production, consumption, and the system context (Chapter 10). Transformative innovations in more "controlled" PCS may be easier, but it may also be less fair. Furthermore, actors whose interests would be seriously threatened by the spread of an innovation can be expected to resist it (Chapter 10).

The knowledge which is brought to bear is both research- and experience-based and can come from almost any part of a PCS. Innovations to link knowledge and action more closely could arise in any part of the PCS, but only some are likely to spread through it and bring more integrated management or influence.

In summary, we propose that actors who are able to bring knowledge and action closer together across disparate pieces of a PCS are effective precisely because their interventions make it possible to influence the evolution of a PCS in a more integrated fashion. To some extent these interconnections may be institutionalized, as for example, through certificates and labels.

Deliberated

Knowledge–action gaps are closed by people working in particular physical or virtual places, or arenas, where knowledge, power and action are frequently simultaneously at play (van Kerkhoff and Lebel 2006). Effective arenas foster more and better quality interactions among actors from different parts of a PCS. Arenas may change the subject and issues addressed, effectively, reframing problems and solutions of sustainability, making them both knowable and actionable.

They do firstly by making and supporting places where learning can take place. Nölting (Chapter 8) for example describes the role of organic fairs on promoting understanding among producers and consumers. Giap et al. (Chapter 7) emphasize the importance of shrimp growers' associations and clubs. But a difficulty in many settings is that small producers can be at a major disadvantage in dealing with complex politics around trade, standards and certification. Their representation at dialogue events and public consultations is often limited in the absence of civil society support.

The three case studies on tourism each argue for greater involvement of local communities in decision-making around tourism developments (Chapters 12 and 13). Taking a PCS perspective suggests all actors have responsibilities and plausible actions they could take that would support sustainability. Bringing various actors involved much more closely together, for example, through committees with broad representation, could help ensure that policies at different levels don't just focus on benefits but also on reducing burdens and risks (Chapter 12). This implies that stakeholders should be defined to include "affected communities" and not just policy-makers, beneficiaries and regulators. Voluntary measures like visitor rules and education for tourists or codes of conduct for service providers will need stakeholder involvement in their design (Chapter 12). At the same time all three studies emphasize the importance of enforcement of existing legislation that protects local environments and health, but which for various reasons are ignored.

Deliberative elements don't necessarily mean consensus-building and negotiation will follow. Indeed creating space or arenas for sharper more radical claims may be crucial for knowledge to inform actions. The vacation rights initiative explored by Maniates (Chapter 2) illustrates the importance of clear, uncompromising messages, in building coalitions and creating public policy pressure. Softer more inclusive approaches did not work.

We propose that actors who create and facilitate arenas in which knowledge and alternative actions can be deliberated and challenged can help transform PCS by reframing and negotiating problem definitions and plausible set of solutions. New understandings and possibilities of action may be institutionalized as standards or agreements to share data or cooperation on R&D.

Validated

Knowledge–action gaps are closed by procedures which lead to progressive improvement. This might be through positive reinforcement like learning or more negative ones like editing out misinformation or bad practices through penalties and sanctions. Monitoring and evaluation of some kind takes places and gives feedback to key actors in the system. We term this third process of closure 'validation'.

Knowledge systems can be full of propaganda and misinformation. Trustworthy sources save time and effort. Authorizing knowledge is an important boundary function taken on by organizations of various forms. One example are shrimp growers associations in Thailand (Lebel et al. 2008). Farmers gain a lot of value from their peers about markets, diseases, prize squeezes and technical matters. They have a seminar – consultant culture, and surprisingly are often more interested in diversification/sustainability of environment then the national research community because their practical experiences underline the importance of water quality and controlling "stress" (Chapter 7).

Tourism development is fraught with schemes of dubious quality which validation could undermine and replace. In Pulau Langkawi tourists have largely ignored a theme park with rice fields, preferring the real to the manufactured experiences (Chapter 11). Over-commodification of culture leads to loss not creation of value. Framing tourism as sun-sand-surf seems to make it sound as if it is endlessly renewable, but each of the studies in this book, point how this rhetorical tactic is misleading.

The interaction between people who know things about sustainability and those able to act on it across a PCS can lead to changes in the goals and criteria used to evaluate and reward performance. More generally they can challenge and sometimes transform the values which seem to underpin unsustainable PCS. An example is the take back your time movement's effort to reframe lower consumption as an opportunity rather than a sacrifice (Chapter 2). Experiences can validate campaign claims.

We propose that actors who validate knowledge and evaluate actions support learning loops which enable PCS to evolve to perform better over time and deal with uncertain and changing circumstances including markets and resource availability. Learning may be institutionalized in good practice guidelines, reporting and auditing procedures. But it also comes with experience or learning by doing.

Reflections

In this final section we reflect on the significance and main limitations of our findings, and make some suggestions on ways forward for scholarship and policy. We draw on the initial summary of different forms of gaps and closures (Table 14.1) and the set of three propositions this led us to in the last section.

Scholarship

Research on the transformative opportunities in knowledge–action links to transform unsustainable PCS is still in its infancy despite at least a decade of efforts to organize research agendas (Stern et al. 1997; Princen et al. 2002; Hertwich 2005; Tukker et al. 2008). Many of the authors in this volume struggled with the combination of inter-disciplinarity and engagement with practice that seems to be required for progress in understanding and influence unsustainable PCS. Substantially more research is needed in several areas.

First, is the question of "*which system?*" The ways in which system boundaries and core elements of a PCS are defined appears to matter a lot for the quality of insights that can be extracted through research and what is salient to practice. Actor identification and mapping appear to be an essential compliment to more conventional focus on material flows, value-added, and step-wise impacts. But attention is also needed to how issues are framed by the actors and alternatives that might be suggested by researchers. In this context exploration of other domains than the relatively restricted set considered in this book is likely to be rewarding. We suggest that much could be gained, for example, from systematic exploration of case studies of transport or mobility systems, housing and shelter provision, and water management.

Second is the question "*whose knowledge?*" Different actors in the PCS have their own unique, but also constrained understanding of places, processes and products. This often does not include much about their environmental or social consequences, locally or more distantly. Only when affected communities, workers and researchers are brought into the picture are these other crucial aspects of sustainability likely to be considered. But that usually raises problems of multiple claims to the truth: whose knowledge is valid, and, how do you combine diverse insights? Power and influence are not easily separable from knowledge claims when it comes to complex problems in PCS. Multi-stakeholder dialogues could be valuable tools in these situations.

Third, is the question of "*how to engage?*" Producers and regulators may call on researchers to carry out new studies or make an assessment of current state of the knowledge to help guide their practices and policies. Consumer organizations call on researchers to help study consumer behavior. Researchers, on the other hand, may demand that consumers or regulators pay attention to their findings about unsustainability and ways that it could be addressed. Our understanding of how individuals and organizations cross the boundaries between normal communities in which they work and seek to influence each others activities are not well understood in the context of PCS. Long commodity chains, for example, pose logistical challenges to direct and frequent engagement unless you are directly part and beneficiary of those transfers like a trader might be.

Fourth, is the question of "*how change is secured?*" Theories about system transformation are still ahead of empirical evidence for many kinds of PCS. A focus on individual and organizational actors that many of the case studies in this volume adopted seems worth pursuing further as it will help addresses the opportunities and limits of agency. It also addresses questions like "*who benefits?*" But it is also important to understand the landscape in which innovations are explored and niches emerged as various elements in a regime must usually co-evolve for major transformation to be achieved (Geels 2005; Schot and Geels 2007). The relationship between social institutions and agency in the longer-term is complex and two-way (Chapter 3). Finally, the role of research-based knowledge in transformative change is understudied as is the knowledge which comes from learning-by-doing. Investing in better understanding of how PCS are changed, and the role new knowledge and innovations had in those changes, will be helpful to making changes in other systems that are unsustainable. A key area is to support and critique activist-oriented initiatives with reflective research on what builds support and secures changes in behavior and practices.

In summary, much more research is needed on how scientific and other forms of knowledge that support sustainability are produced, shared and used in PCS. These analyses will compliment and extend the much larger volume of sectoral-, place-, technology- and consumer-focused research already in hand.

Policy

The case studies in this volume and this synthesis imply knowledge–action gaps in PCS are frequent and benefit from explicit efforts at closure (Table 14.1). Several findings stand-out as broad significance for policy.

First, adopt a systems view (Lebel and Lorek 2008). Go beyond the conventional narrow focus on single firms, processing steps, actor groups, or sectors as the starting points. The PCS perspective immediately highlights another set of leverage points in interactions among nodes along a commodity chain and other actors attempting to influence or otherwise affected by those chains. Research and action agendas can focus on those links, improving their information content, as well as mutual understanding among different actors involved in these relationships.

Second, jointly consider the production- and consumption-side issues and perspectives in each relationship (Princen et al. 2002). Knowledge–action gaps may be closeable from both demand and supply-side considerations. Don't blindly assume producer responsibility or consumer sovereignty or full and transparent knowledge, but rather expect a need for informed dialogues and negotiations.

Third, be skeptical of information-only based approaches targeted at one group, producers or consumers. Lack of integration across PCS means changes are unlikely to spread and may even be compensated for by changes in practices in other nodes with result of no net benefit. Regulatory and incentive systems often need to align and reinforce knowledge elements. Action does not follow from knowledge, nor does knowledge necessarily come from learning by doing, without explicit efforts to make it happen. Incentives and interests often need to also be aligned and shown to be so.

Fourth, don't discount the importance of policy and looming standards or regulations. The highest priority actions to narrow gaps between knowledge and action might be on the policy front. This was pointed out in the case study on tourism in Goa which identified a strong need for better enforcement of existing environmental legislation and plans (Chapter 13). Likewise the German 'feed-in-law' for electricity from renewable sources proved to be more relevant than disclosure information to consumers (Chapter 5). Innovations in business knowledge and practice are often triggered by stringent environmental policy; a large part of "voluntary" actions of companies and environmental product innovations are driven either by existing or anticipated regulatory actions. Framing policies should stimulate and coordinate the transition process without suffocating innovation (Rehfeld et al. 2007; ASCEE team 2008; Frondel et al. 2008).

The landscape in which innovations in PCS unfold is highly dynamic (Chapter 1). The kinds of strategies needed must also vary with time. The strongest examples are from shifts in global economic system including trade, finance and investment. Instability in the global economic system creates both opportunities and threats to efforts to make PCS more sustainable. On the one hand, recessions, can lead to individuals, households and firms re-examining their consumption patterns in efforts to reduce expenditure. Consumers are often looking for ways to make cuts without reducing quality of life. Producers may also turn to innovations that save on material or energy inputs or otherwise improve competitiveness. On the other hand, reduced availability of finance and lower profits can also mean cuts to longer-term research and development projects. Instabilities can also lead to re-examination by firms and governments of trade relations and policies with some countries opting to further globalization with less constraints on exchange whereas others respond with subsidies, tariffs and other instruments to try and protect own industries.

The policies needed in the pursuit of sustainability for different PCS are diverse (Martens and Spaargaren 2005; Fuchs and Lorek 2005) but one common feature is that they need to address knowledge–action links carefully or they are likely to fail.

Conclusions

Transforming unsustainable production–consumption systems (PCS) is crucial to pursuits of sustainability. The challenge is frequently not a problem of insufficient knowledge, but of failures to close knowledge–action gaps in which interests and power relations play significant roles.

Across diverse range of initiatives, including a set explored in detail in the chapters of this book, some hints of what it takes to succeed are emerging.

First, actors who target their efforts at bringing knowledge and action closer together across disparate pieces of a PCS are effective precisely because their interventions make it possible to influence the evolution of a PCS in a more integrated fashion. Links once created may be institutionalized, for example, as traceability mechanisms.

Second, actors who create and facilitate arenas in which knowledge and alternative actions can be deliberated can help transform PCS by reframing and negotiating problem definitions and plausible set of solutions. New understandings and possibilities of action may be institutionalized as standards or agreements to share data or cooperation on research and development.

Third, actors who validate knowledge and evaluate actions support learning loops which enable PCS to evolve to perform better over time and deal with uncertain and changing circumstances including markets and resource availability.

Learning may be institutionalized in good practice guidelines, reporting and auditing procedures.

Effective change agents create and support knowledge–action arenas, look for ways to reframe value changes as opportunities, and support institutional changes that will secure past, and enable future, gains.

Acknowledgements We are grateful to friends and colleagues that participated in the virtual and physical meetings of the SPACES working group for their inspirations that led to this collected volume and influenced our thinking in many ways. We are grateful to the David and Lucille Packard Foundation for their financial support.

References

ASCEE team (2008) Policy instruments to promote sustainable consumption. EU FP 6 Project ASCEE – Assessing the potential of various instruments for sustainable consumption practices and greening of the market. Brussels/Heidelberg/Oslo

Bailey C (1988) The social consequences of tropical shrimp mariculture development. Ocean Shoreline Manag 11:31–44

Bilharz M, Lorek S, Schmitt K (2008) "Key points" of sustainable consumption. In: Geerken T, Tukker A, Vezzoli C, Ceschin F (eds) Sustainable consumption and production: framework for action. Brussels, Belgium, pp 10–11, March 2008 TNO, Delft, Netherlands

Burgess J, Harrison C, Filius P (1998) Environmental communication and the cultural politics of environmental citizenship. Environ Plann A 30:1445–1460

Cash D, Clark WC, Alcock F, Dickson NM, Eckley N, Guston DH, Jager J, Mitchell RB (2003) Knowledge systems for sustainable development. PNAS 100:8086–8091

Connolly J, Prothero A (2003) Sustainable consumption: consumption, consumers and the commodity discourse. Consumpt Mark Cult 6:275–291

de Zoysa U (2005) Asian review on sustainable consumption. Centre for Environment & Development, Malaysia

FAO, NACA, UNEP, WB, WWF (2006) The international principles for responsible shrimp farming: shrimp farming and the environment. Network of Aquaculture Centres in Asia-Pacific (NACA), Bangkok

Frondel M, Horbach J, Rennings K (2008) What triggers environmental management and innovation?–empirical evidence for Germany. Ecol Econ 66:153–160

Fuchs DA, Lorek S (2005) Sustainable consumption governance: a history of promises and failures. J Consum Policy 28:261–288

Geels FW (2005) Processes and patterns in transitions and system innovations: refining the co-evolutionary multi-level perspective. Technol Forecast Soc Change 72:681–696

Goss J, Burch D, Rickson RE (2000) Agri-food restructuring and third world transnational: Thailand, the CP Group and the global shrimp industry. World Dev 28:513–530

Green K, Foster C (2005) Give peas a chance: transformations in food consumption and production systems. Technol Forecast Soc Change 72:663–679

Guston DH (2001) Boundary organizations in environmental policy and science: an introduction. Sci Technol Hum Values 26:399–408

Hertwich EG (2005) Life cycle approaches to sustainable consumption: a critical review. Environ Sci Technol 39:4673–4684

Hobson K (2003) Thinking habits into action: the role of knowledge and process in questioning household consumption practices. Local Environ 8:95–112

Jackson T (2005) Live better by consuming less? Is there a "double dividend" in sustainable consumption. J Ind Ecol 9:19–36

Kates RW, Parris TM, Leiserowitz AA (2005) What is sustainable development? Goals, indicators, values and practice. Environment 47:8–21

Knight A (2004) Sustainable consumption: the retailing paradox. Consum Policy Rev 14:113–115

Lebel L, Lebel P, Garden P, Giap DH, Khrutmuang S, Nakayama S (2008) Places, chains and plates: governing transitions in the shrimp aquaculture production-consumption system. Globalizations 5:211–226

Lebel L, Lorek S (2008) Enabling sustainable production-consumption systems. Ann Rev Environ Res 33:241–275

Martens S, Spaargaren G (2005) The politics of sustainable consumption: the case of the Netherlands. Sustain: Sci Pract Policy 1:1–14

Princen T, Maniates M, Conca K (eds) (2002) Confronting consumption. MIT Press, Cambridge, MA

Raynolds LT (2004) The globalization of organic agro-food networks. World Dev 32:725–743

Rehfeld KM, Rennings K, Ziegler A (2007) Integrated product policy and environmental product innovations: an empirical analysis. Ecol Econ 61:91–100

Renard M-C (2003) Fair trade: quality, market and conventions. J Rural Stud 19:87–96

Rorty R (1979) Philosophy and the mirror of nature. Princeton University Press, Princeton, NJ

Schor J (2005) Sustainable consumption and worktime reduction. J Ind Ecol 9:37–50

Schot J, Geels FW (2007) Niches in evolutionary theories of technical change: a critical survey of the literature. J Evol Econ 17:605–622

Seyfang G (2004) Consuming values and contested cultures: a critical analysis of the UK strategy for sustainable consumption and production. Rev Soc Econ 62:323–338

Stern PC, Dietz T, Ruttan VW, Scolow RH, Sweeney JL (eds) (1997) Environmentally significant consumption: research directions. National Academy Press, Washington, DC

Tukker A, Emmert S, Charter M, Vezzoli C, Sto E, Andersen M, Geerken T, Tischner U, Lahlou S (2008) Fostering change to sustainable consumption and production: an evidence based review. J Cleaner Prod 16:1218–1225

van Kerkhoff L (2005) Integrated research: concepts of connection in environmental science and policy. Environ Sci Policy 8:439–463

van Kerkhoff L, Lebel L (2006) Linking knowledge and action for sustainable development. Ann Rev Environ Res 31:445–477

Vandergeest P (2007) Certification and communities: alternatives for regulating the environmental and social impacts of shrimp farming. World Dev 35:1152–1171

Index

T

LaVergne, TN USA
09 May 2010
182044LV00001B/10/P

9 789048 130894